A-4 / V-2 Rocket

Instruction Manual
(in English)

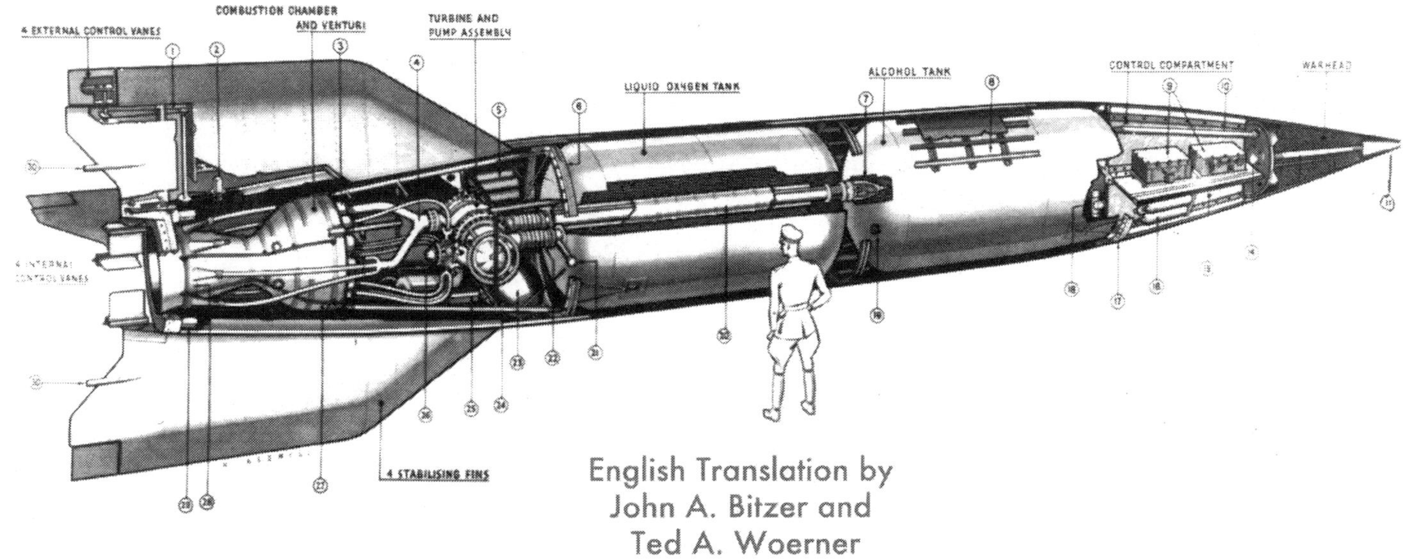

English Translation by
John A. Bitzer and
Ted A. Woerner

ISBN #978-1-937684-76-1

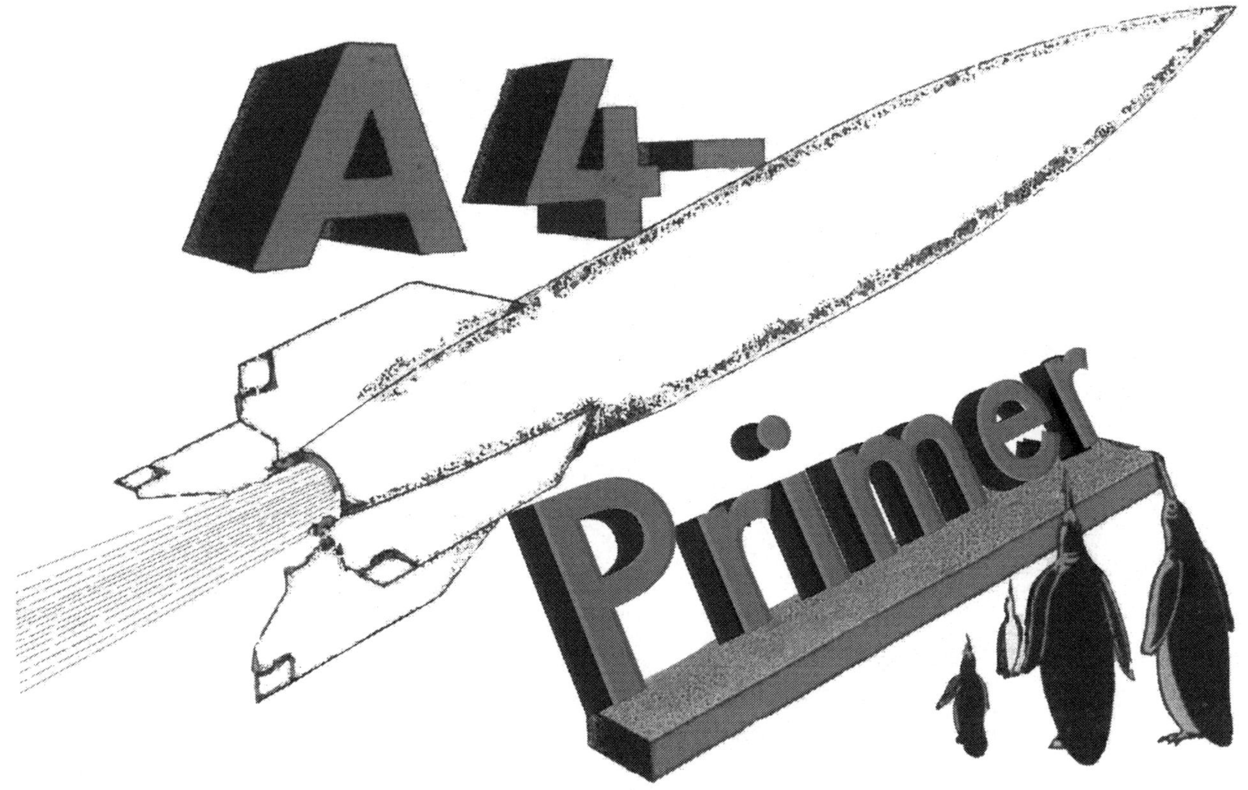

Originally published by the
Army Ballistic Missile Agency
Redstone Arsenal
1957

FOREWORD

This translation of the A4 Fibel (Primer) presents the German version of a technical manual for the V-2 missile, an operator's guide which bridged the gap between technical laboratory reports and the field soldier who used the equipment. The illustrated approach, with points vitalized by cartoons, might be considered a basic interpretation of highly technical matter for the benefit of the man on the firing line.

A wide gulf lies, and will always lie, between the research and development laboratory and the field soldier who is called upon to use the completed and standardized esult of that laboratory's research. Field manuals and technical manuals, almost without exception, are dry and uninteresing in their methods of presentation, and the soldier who seeks simplicity finds himself bogged down in a mass of unnecessary detail and technical jargon which have been written with the assumption that " the man in the field is throughly familiar with the methods and techniques reported.

Those who have undertaken the translation and reprinting of the A4 Fibel have felt that the German originators at least approached the solution of the problem of the overly complex manual and that those who read it might recognize the need for directness and simplicity in the preparation of similar documents which might be used by persons whose education is not that of the technical laboratory.

This reprint of the A4 Primer preserves the page for page format of the original text. The editing procedure adopted was to assemble, translate, and present the data in as identical as the primary sources would allow. Since the A4 handbook was issued under Peenemunde security regulation, the initial German publication of 150 copies was restricted to field troops. Inumerable conflicting line-lengths were expected and were encountered but the problem was solved by use of an 8 X 10-1/2 page instead of the smaller pocket size of the original. Considerable effort has been made to achieve accuracy and completeness in translation; accuracy in the sense of avoiding technical errors and completeness by including details of the original edition, including all illustrations, arrangement, and reproduction of the cover page. It should be remembered that the significance of the jingle-mottos frequently depends upon German idiom which unfortunately is lost in translation.

It is felt this A4 Fibel will prove of value as reference for the general researcher who seeks background information about the state of the art, as well as those who desire to be informed on project or program history as presented by Peenemunde Research Institute. The A4 Fibel was translated by Sp Ted Woerner, Ordnance Corps; editorial supervision was under the direction of John A. Bitzer; publication was by the Reports and Publications Section, Development Operations Division, Army Ballistic Missile Agency, Redstone Arsenal, Alabama.

-: JOHN A. BITZER

TABLE OF CONTENTS

1. First Print (Check Number 1 to 150) according to the situation as of 1. 7. 1944

O.U., 4 August 1944

I authorize the A4-Fibel

Everybody listen up!

You, dear reader, find here
the new Fibel for the A4.
The boring stuff, with premeditation

in a easy form presented,
makes it more pleasant for you,
it should be evident and convenient
to go into your flesh and blood
if it did not happened yet.
So far everything is good and beautiful,
but, please, one cannot be overlooked:

The whole stuff about the A4 is top secret, just remember this!

Who speaks about it, perpetrates treason,
it hurts you and the state!
Just remember one thing,
Do not take part in any debate!
Make yourself not smarter then you are,
because it is not of your own knowledge.
Stay away from discussions
you only get in trouble!
And you get ask by an outsider
an informer or a smart ass,
then just go into full retreat,
Otherwise you could loose your head!
Put on your stupidest face and say:
"My name is rabbit, I do not know nothing!"

You live on this earth planet
at the age of long distance rockets.
The sky-ship in space -
A work of peace and a human dream -
Greeted the new century!
Now it means to master a new weapon:
As a secret equipment still unknown,
here known as A4 apparatus.

You belong to the FR - Force,

which operates the A4 long distance rocket. You are also involved with the projectile, which is larger and flies further then all other known projectiles.

The A4's powerful blast effect, cannot up to now be matched by any other known projectile or bomb.

At the launch preparations you are working hand in hand with your comrades. It lays in your hands, to quickly prepare the A4 for launching and for reaching the target,

The projectile maintenance is part off it. Use your free time.

Only with a well planned maintenance program of all the equipment and tools, could a safe preparation for the next launch be guaranteed.

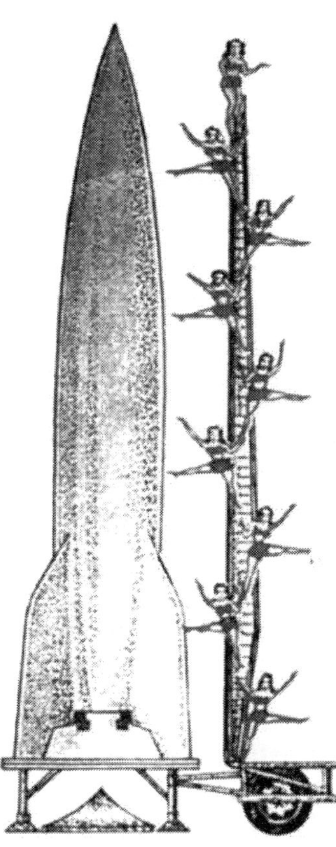

Moral: To be perfect is not easy
take care of every little detail it will lead to perfection

Faster then the speed of sound

14 m long is the A4, it weights 12.5 t and it reaches a range of 300 km. The speed when it hits the surface is 800 m per second. The A4 also flies faster than the speed of sound. You cannot here the arrival; you only hear a gigantic explosion, but then it is too late. The top speed of the A4 is reached 1 m after launch. The speed is double as when it hits the target, around 1500 m per second, this is about 5400 km per hour, which is about 8 times faster then the fastest airplane.

Only 5 minutes passes between launch and hitting the target.

But in the 5 minutes everything has to work. All parts of the A4 have to be checked and repaired before a launch, so that shot will hit the target. Only a small matter could be the reason for a failure..

<u>Remember:</u> Every launchfailure helps the enemy and hurts us through the loss of valuable materials, and also endangers you and your comrades.

Moral: The rocket takes it badly
if you do not look into the fibel.
But it gives the enemy trouble after every successful shot.

MOTTO: You will save yourself
much trouble in life,
If you always know
what you should know.

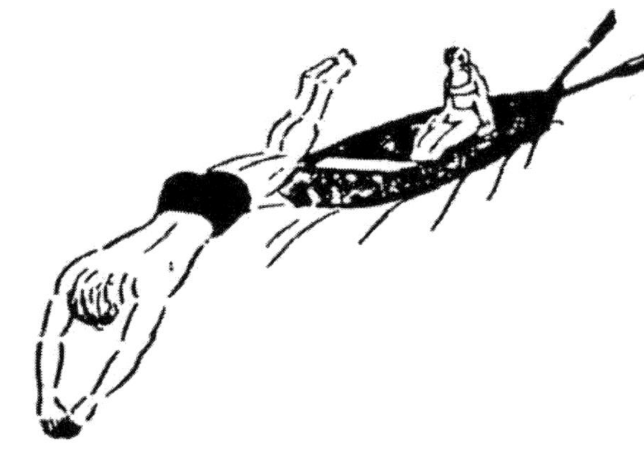

50 WORDS OF TECHNIQUE IN REGARD TO THE A-4

Jet Propulsion:

man jumps into a river from the stern of a boat. The boat will move forward. This is caused by the backward pressure. The same principle is used for the propulsion of the A-4.

The combustion unit is installed at the rear. High pressure combustion gases flow from it and drive the A-4 forward.

The Thrust is the force produced by propulsion.

The gases of combustion are formed by burning two liquids called A and B substance. More than 8000 liters of both substances are needed for filling up the A-4. They will burn up in one minute.

A-Substance aids combustion. It is unbelievably cold (183 degrees). You have to be real careful in handling it, or you will suffer freezer burns. Wear at least asbestos gloves A-substance is inflammable.

B-Substance is the fuel that burns up. It is explosive and poisenous if used internally.

A-Substance and B-Substance are fed to the combustion unit separately.

Jets spray the two propellants. In the head of the combustion unit you will find many hundred jets. These jets should not be dirty, otherwise they will cause failure at launching.

The Pressure in the Combustion Unit is 15 atu. This is necessary in order to facilitate a fast exit of the gases of combustion from the combustion unit.

A-Substance and B-Substance Pumps. Against a pressure of 15 atu, A and B Substance has to be pumped into the Combustion unit. Therefore we need two rotary pumps, one A-Substance pump (1) and one B-Substance pump (2).

One Steam Turbine (3) drives the two rotary pumps. It is located between the pump.

At the turbine-driven auxiliary the steam turbine and the two pumps are combined.

One steam generating plant is utilized to chemically generate steam for the turbine. Here steam must be generated within a fraction of a second.

The Steam Generator generates steam by mixing two substances, T-Substance and Z-Substance together.

The T-Substance generates over-heated steam. T-Substance affects the skin like an acid and inflames clothing. Wear protective clothing.

Z-Substance acts as a decomposing agent.

The Steam Pressure amounts to approximately 32 atu. T-Substance and Z-Substance must be fed into the steam generator under this pressure.

P-Substance (compressed air or nitrogen) is carried in the A-4 in pressure-bottles of below 200 atu.

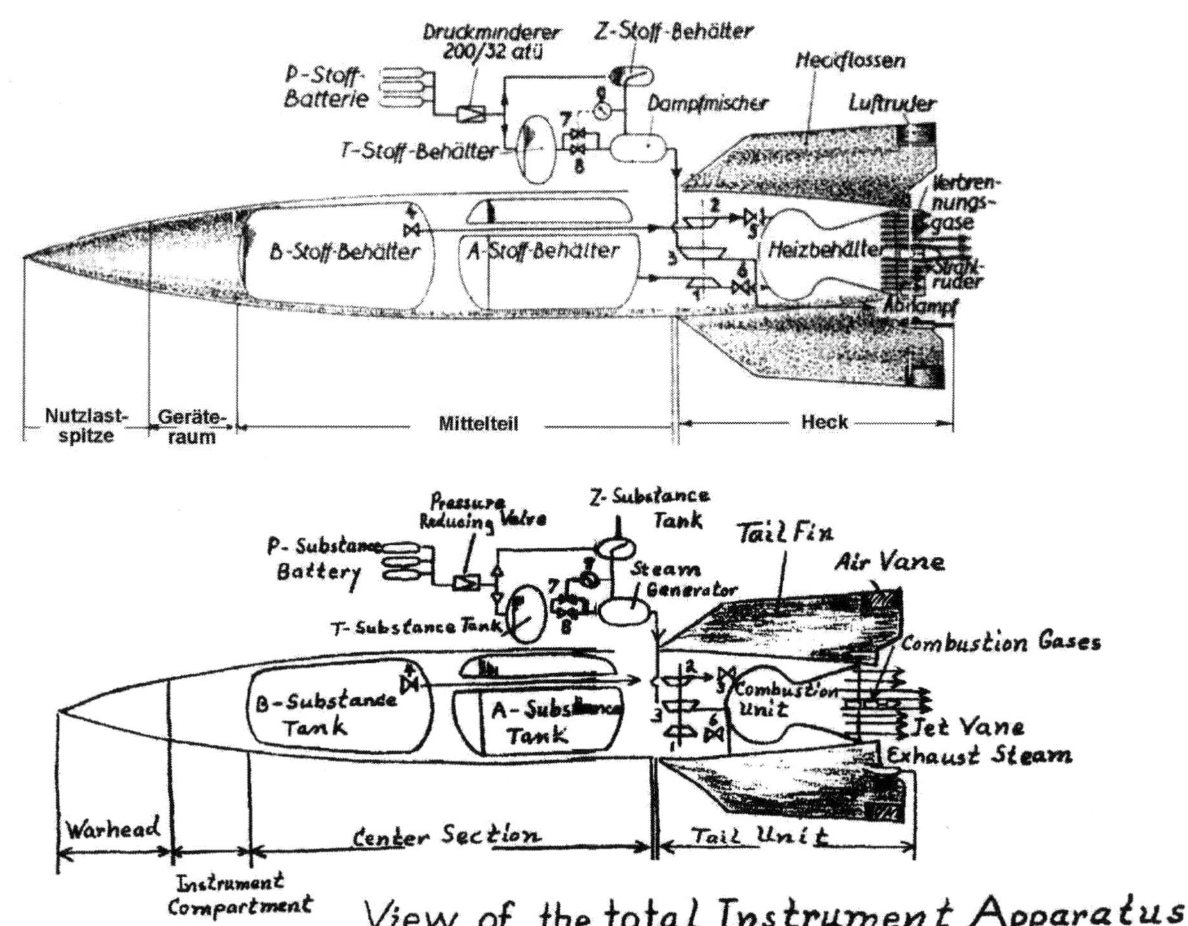

View of the total Instrument Apparatus.

The Pressure Reducer — Here the pressure of the P-Substance (200atu) is reduced to approximately 32 atu. The steam reducer has to be adjusted perfectly according to the pressure readings on the shipping document.

The B-Substance Pilot Valve (4) is located at the bottom of the B-Substance tank; it shuts off the drain line.

The B-Substance Main Valve (5) is installed in the head of the combustion unit. It regulates the B-Substance flow to the spray jets.

The A-Substance Main Valve (6) is located at the pressure nozzle of the A-Substance pump. It blocks the A-Substance supply to the spray jets of the combustion units.

The T-Substance Valves — T-Substance will flow through the 8-ton pressure valve (7) and the 25-ton valve (8) to the steam generator in such a quantity that the produced thrust of the A-4 will amount to approximately 8 tons, or as the case may be 25 tons.

A Pressure Contact (9) causes the valves (7) and (8) to open only then when a little of the Z-Substance has already entered the steam generator. Otherwise an explosion could occur when T-Substance and Z-Substance meet.

The Main Parts of the A-4 are the warhead, the instrument compartment, the middle part, and tail unit. Instrument compartment, middle part, and tail unit are made accessible through shutters.

The Shutter Diagram will show you where the individual shutters are located.

The war head contains the explosive.

Instrument compartment with Sections I and IV carries batteries, steering equipment, and radio equipment.

The Middle Part — contains mainly the A-Substance storage tank and the B-Substance storage tank.

The Tail unit — contains the Thrust-Unit. (The drive)

The Four Fins — located on the tail unit prevents the A-4 from turning over during flight. They are numbered 1 to 4.

The Thrust Unit consits mainly of a combustion unit, a turbo-generator and steam generator. It converts the energy of fuel into push power, (or thrust)

On the FR-Carrier the A-4 will be mounted horizontally and is transported into the firing position in that way. It is erected by utilizing a crane-device and is ercted on a launching platform in a vertical position.

The Fueling with A-Substance, B-Substance, T-Substance, Z-Substance, and with P-Substance will be done when the device is erected.

Steering: when fired the A-4 will take off vertically. Through installed automatic control it is held on the prescribed course and is steered in such a way that upon termination of propulsion the missile will be at a certain angle to the horizontal.

The Rudders: The A-4 is steered by the rudders. Four jet-rudders are located in the stream of the gases emerging at the tail unit and four air-rudders operate at the outer edge of the Tail-fin.

Propulsion: As long as combustion gases flow from the A-4 the missile will be driven forward. Air speed increases during propulsion.

Cut Off: Approximately one minute after firing, the fuel supply of the combustion unit will be cut off. After that the A-4 will fly like a normal shell that has left the gun barrel. Two methods may be emplozed for cut off.

1. Cut off from radio signal given from the ground.
2. Cut off via the installed impulse device in the missile itself.

The Range: The firing is regulated by an earlier or later cut-Off of the fuel supply. Propulsion is therefore cut off at a lower or higher velocity. The angle of take-off is always the same.

The Direction: The direction in which the missile will fly is regulated by the turning of the launching platform. The steering operates in such a way that the A-4 flies in the direction of fin Number 1.

Due to the Localizer Beam Method the deflection dispersion may be reduced by radio.

Electric Batteries: which are aboard, supply the A-4 with the necessary power while flying.

The Launching Site: The A-4 is launched from the launching site.

The Ground Device: In conclusion the A-4 must be adjusted and tested. For this and for launching the ground installation is utilized, consisting of various types of equipment.

The Power Supply Trailer supplies the electricity necessary for testing the A-4.

The Compressor Unit is available for producing the necessary compressed air. (P-Substance)

The Launch Conntrol Car has panels installed through which the preparatery and firing process may be tested.

The Combustion Cut-Off Site harbours the apparatus utilized for disconnecting the driving mechanism via radio. This equipment is located 0 to 14 kilometers, usually 7 to 9 km behind the firing site.

The Guide Beam Transmitter is set up approximately 10 to 16 km, behind the launching site. The Guidebeam transmitter afords a more accurate lateral guidance of the A-4.

The X-Time Chart governs synchronization and flow of the individual work processes. You will locate it in the appendix. The X-time chart gives you enough time to complete your work satisfactorily. You will figure the time backwards. At X-100 it will be 100 minutes to launching.

MORAL: A theory is a much hated thing and yet, nothing can be accomplished without it.

THE SHUTTER PLAN

(1) Hatch for A-Substance and T-Substance filling

(2) Hatch for Z filling

(3) Hatch for pressure reducing valve /4) Hatch for Z-Bleeder

(5) Hatch for pressure content and T-draining valve

(6) Hatch for Z-draining valve

(7) Hatch for secondary distributor

(8) Hatch for turning the pumps
and removal of safety regulator cut-out

(9) Manhole

(10) Loading hatch

(11) Serve hatch unit

(12) Spring hatch for magnetic plugs

(13) Ventilation hatch

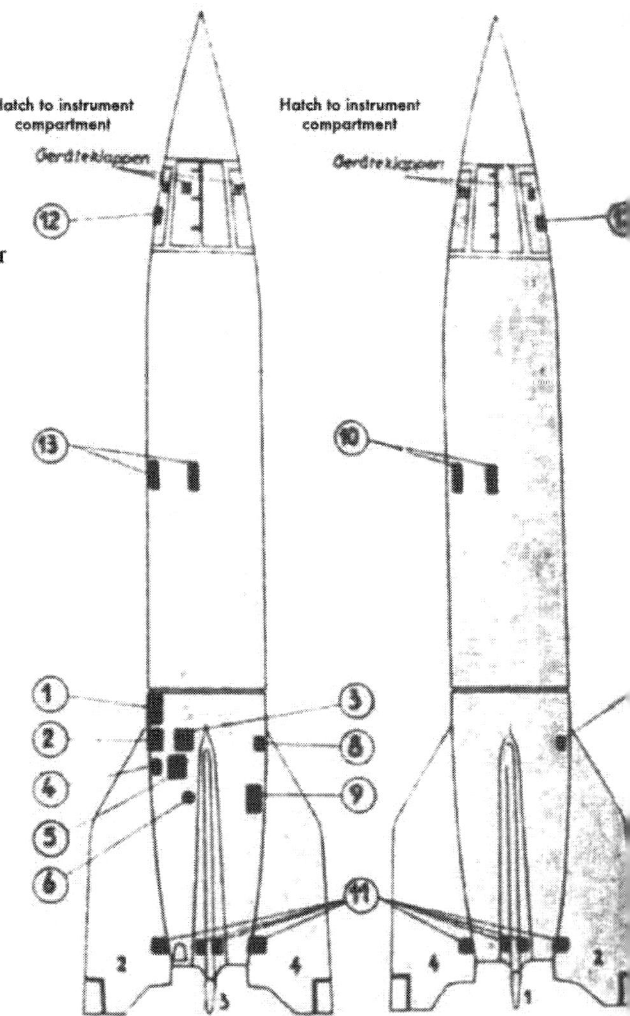

IN THE FIRING POSITION

At the firing site the A-4 placed into a vertical position, it is raised for launching, tested, fueled, and then finally launched. Most of your work is done at the launching site.

THE MEASURING TROOP

The rifle man has it relatively simple, he sees the target that he wants to hit. But you will shoot at a target that cannot be seen. In order to determine distance and direction certain measurements and calculations are necessary. This work has already been done so that you will be in possession of the results once you move into the firing position.

AIMING IS YOUR PRIMARY JOB

You will therefore perform surveying and adjustment jobs.

Accuracy on your part will determine the accuracy of fire.

MOTTO:
He who does everything the way it should be done
Must be a living magican.

TITLE: PREPARING THE POSITION

Your friend has offered you his summer home to live in, due to the fact that your apartment was damaged. You intend to furnish the house with your furniture. In order to save time and prevent confusion you must first measure the furniture and the rooms of the house. The some holds true for the A-4. Here the launching site had also been surveyed prior to your moving in. All the important data is given in the launching site manual.

After your detachment has moved into the launching site, locate the measured points. In order to be sure that they are still correct you must check and re-adjust them. For it is possible that a measured point has been moved or removed maliciously. In such a case you must restore the point.

This work must be completed prior to moving the A-4 into firing position.

Otherwise -- it cannot be moved into the correct position.

After the A-4 has been erected you will make adjustments, that is:

1. Turn it in a vertical position,

Otherwise - the Electro-Detachment will not be able to get the steering mechanism ready for firing.

2. Fin 1 must be turned in the direction of fire,

Otherwise -- it will fly in the wrong direction,

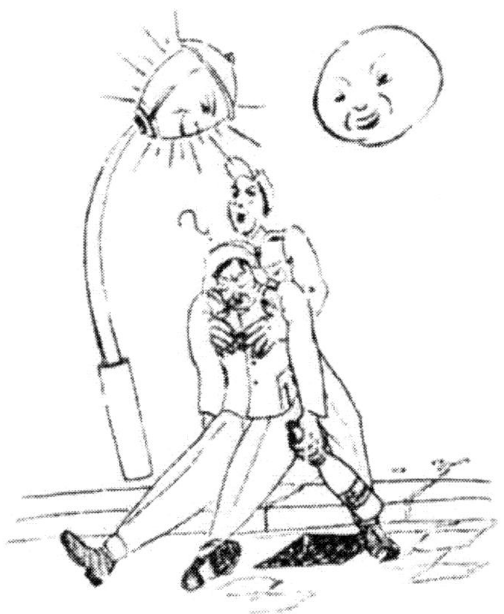

PLACING THE MISSILE IN A VERTICAL POSITION

For placing the missile in a vertical position two aiming-circle collomaters of 12m are necessary, of which, however, you only need the grid plate of the telescopic sight. An aiming-circle collimater is an optical device with which it is possible to lay two optical axis reciprocally and this more accurate than with an aiming-circle.

(1) If possible set up an aiming-circle collimater 12m (Rkr KI), in the direction of fire, approximately 50 meters in front or behind the A-4 and level it.

(2) Do the same with the second circle aiming collimater (Rkr KM) and this vertically, if possible.

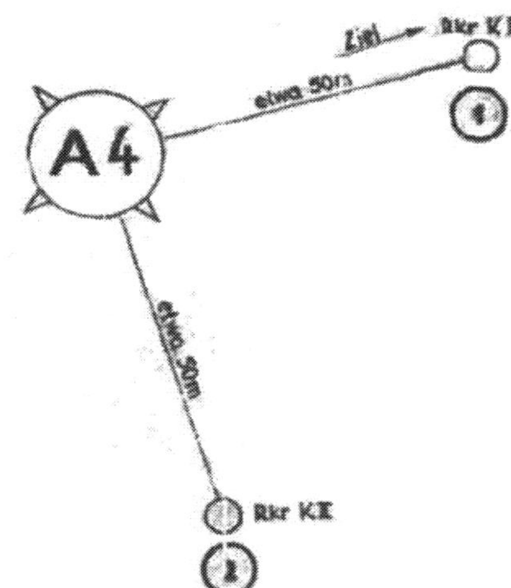

(3) Adjust the telescopic sight of the first aiming-circle collimater at any azimuth reading, by turning the outer part of the circle toward the A-4 in such a way that its left and right edge above the fins on the grid plate are spaced equally from the center of the cross hairs.

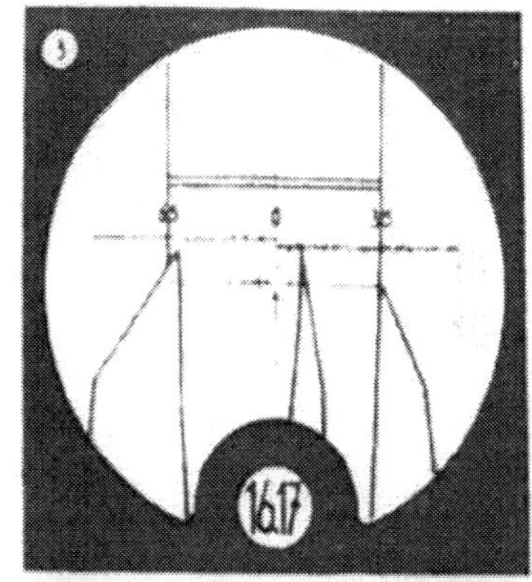

(4) Tilt the telescopic sight by turning the elevating gear until the nose of the A-4 moves into your sight.

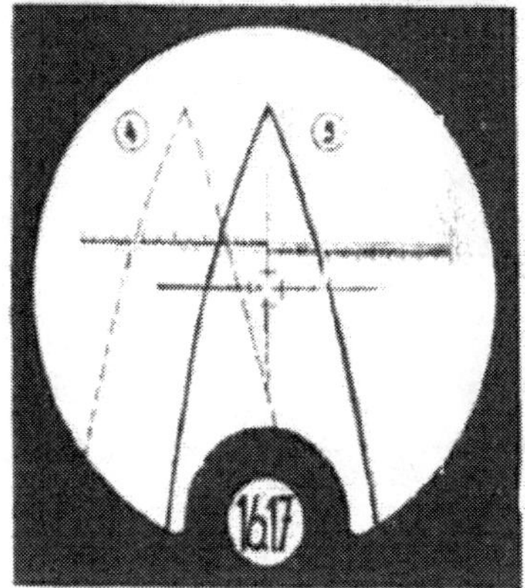

(5) Direct the men of the trailer-crew until, by adjusting the spindles at the launching platform, the center line in the telescopic sight is covered by the nose of the A-4. Do the same from the second aiming-circle collimater (Rkr KM). Repeat the process from Rkr Kl and II.

PREPARING THE POSITION TO THE AZIMUTH OF THE TARGET

When preparing the position to the azimuth of the target you must set Fin # 1 of the A-4 exactly in the direction of fire. It is assumed that the launching site has been surveyed.

There are two surveying points that are especially important:

1. A point directly in the launching site for which the final azimuth of the target and the distance to the target has been calculated.

2. Close to the launching site the aiming point Rl hatter has been chosen in such a way that the control points and possibly the launching sites can be observed.

The launching site manual contains information on the position of the points as well as the aximuth of the target and the control points.

order to make adjustments the following is required:

ng-circle collimaters 12 meters and one dial sight ollimater grid-plate and one frame.

(1) Install the panoramic telescope (Rbf) with frame in control axis #1, #2, and #4. Clamp the movable arm of the frame.

Rkr. K.II
3
R.II
1 Rbf.

(2) Place an aiming-circle collimator 12m (Rkr K II), not to exceed a distance of 12m from the K-4, on an arbitrary point—designated as R II— from which the Rbf on the A-4 and the Rkr K I is visible.

If such a point R II cannot be found then place an additional aiming-circle collimator between R I and R II so that both can be observed.

Remember: The Rbf must remain visible from R II even when turning the A-4 in the direction of fire.

Your Job on the aiming-circle collimator I

(1) By utilizing Rkr K I sight the requested aiming point — for example the church in A-town- by utilizing a requested number-in this case for excample 2573.

(2) Turn the upper part of Rkr K I on point R II by depressing the plunger notch and let it lock. Read-off the excact number - for excample 4829,5. Relay this number to your comrad at the R II.

Your job on the aiming-circle collimator II

Obtain a bearing from Rkr K II to point R I by utilizing the given number. New Rkr K II and K I are parallel.

Turn the upper part of the Rkr K II, by depressing the plunger notch until the Rbf on the A-4 is visible on the picture screen.

Latch into the next notch. You wil call out the aximuth reading to your comrade at the Rbf-always the number on the right side of the found sector—as a hundred number, for excample 3800.

Adjust the collimator periscope on the Rkr K II in such a way that the collimator signal plate is illuminated as much as possible. Should it be dark then attach an electric light.

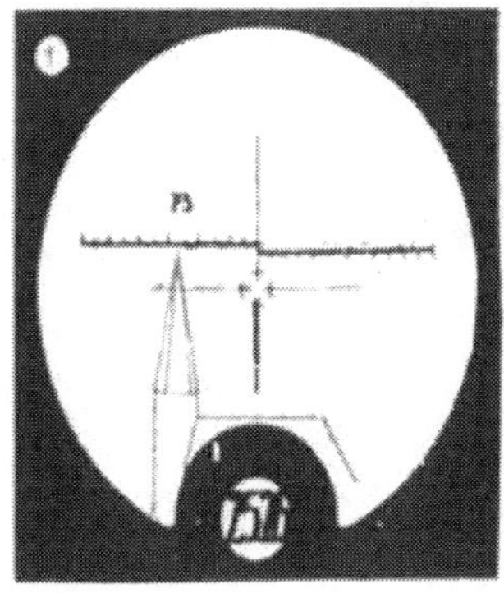

Your Job on the Dial Sight

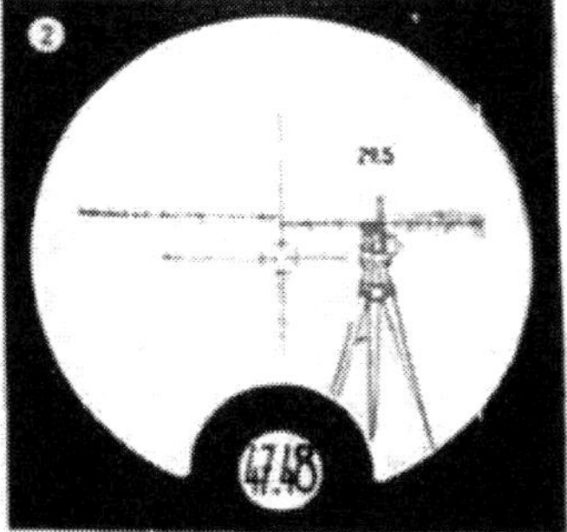

Read off the angle of deviation supplied with a plus or minus sign and indicated on air-rudder #4.

Add the reading to the given azimuth reading 3800 if the sign is + and subtract if the sign is -.

Utilizing this number, make the adjustment on the dial sight.

Direct the men of the trailer-crew at the ratchets of the launching platform in such a way that they will turn the A-4 on the turret ring until the collimator signs, which you will observe in the collimater by utilizing the dial sight, are situated above the same signs you see in the dial sight. For adjusting the field of vision utilize the graduation on the head of the panoramic telescope.

The A-4 is therefore parallel with the aiming circle collimator and points in the direction of the target.

MORAL:

Only then when the A-4 is in a vertical position and when the missile has been turned in such a way that Fin #1 points towards the enemy will the missile be sure to reach its destination.

You will transport the A-4 from the field depot to the launching site over roads and narrow farmways, through narrow curves and low underpasses.

It is up to you whether the missile is delivered without defect or whether it will be damaged through transport. If you damage it you don't have to deliver it. Just take it right back to the field depot.

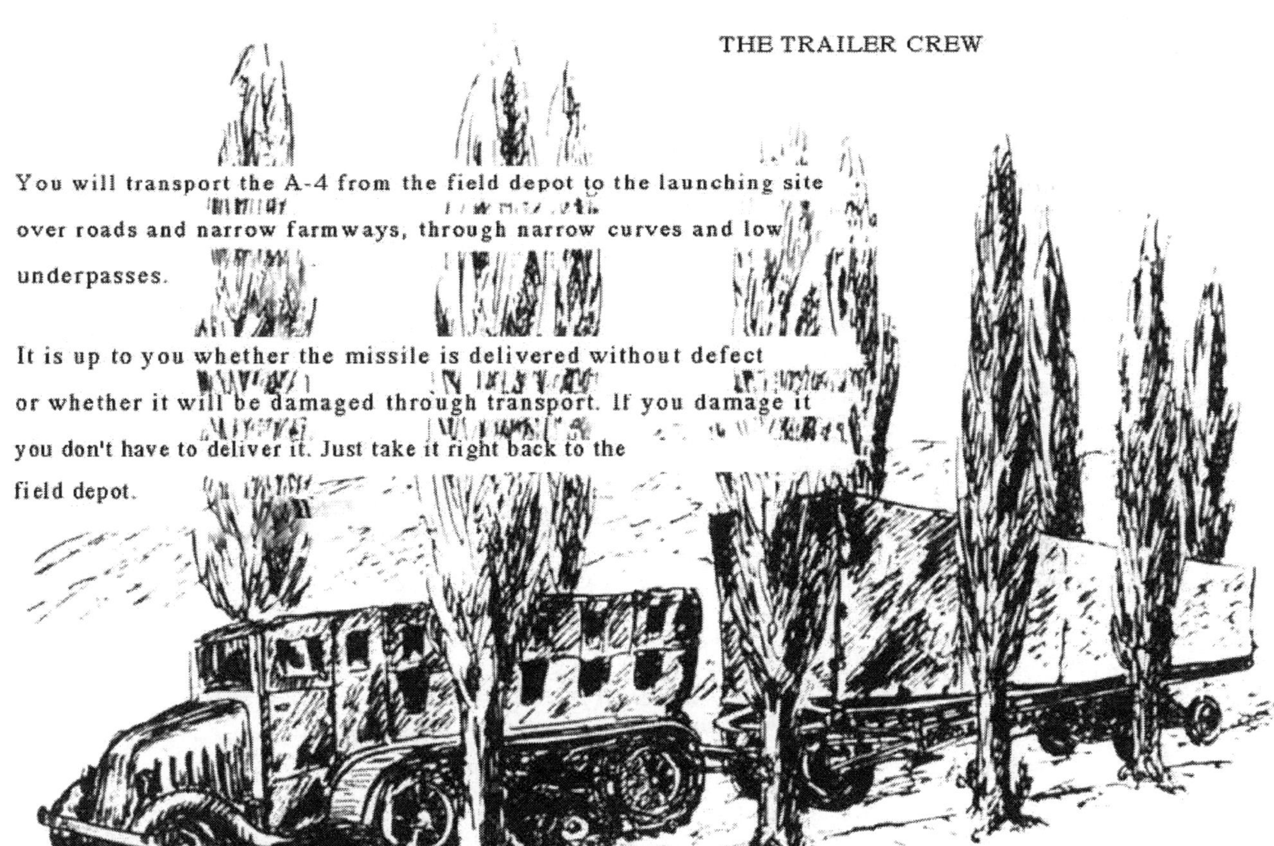

MOTTO: With 80 or 100 km an hour
You make no speed with the A-4.

DRIVING. BUT WITH SENSE

Caracciola drove over the Avus at a speed of 200 km an hour. When going into the curve of the Nurburgring he had to drive slower. But he was driving a racing car and not an A-4.

Like an egg, the many sensitive parts of the A-4 are covered by a shell.

Therefore— you must handle the A-4 in much the same way you handle the egg you bring home when on furlough.

Drive slowly, and carefully and if possible on good roads.

Otherwise— screws will loosen and protruding parts such as Fins will bend and the propulsion unit crew will waste valuable time with repais.

Good Roads: Drive with a speed not over 40 kilometers per hour.

For Cross Country and Field Paths: Drive according to the circumstances.

Curves: Watch that the tires don't rub against the frame.
Otherwise-the tires will be damaged and the outrigger axles warped!
Smallest half curve radius is 10.7 meters.
In the toghter curves you can turn the rear axles separatly.
The hand turnig devoce should be attached for this.

Underpasses: Must be 4 meters high, if you drive with the camoflage frame they must be 4.20 m high.

Otherwise—you will damage the fins!

Dust and Dampness are enemies of the A-4, for dust clogs the propellant jets in the combustion unit, causing the missile not to take off properly. Dampness settles in the valves causing them to freeze up at the low temperature of the A-Substance.

For protection the tail of the A-4 is covered with a tarpaulin. The loops must be tied from top to bottom.

Watch that the loops and zippers of the tarpaulin are closed and that the tarpaulin is not damaged.

Otherwise--your trip will be useless!

Trees: Should you pass trees too closely the branches will tear holes in your camoflage tarpaulin and damage the A-4. Remember, your trailer is 2.87 meters wide.

At The Launching Site:

park at the designated spot. Unhook your tractor and should it not be needed any more then park it in the close-limber position.

Now Remove the Tail Tarpaulin!

Sequence of operation:

(1) Open

(2) Roll off

(3) Pull out

MORAL: Where prudence and forethought are combined
The trip will surely be a success.

MOTTO:

Momentum does not always accomplish
What constant persistence may.

ERECTION OF THE A-4

To do a handstand in gymnastics requires a certain amount of momentum. A well-trained athlete applies the correct amount of momentum so that he will not fail.

If you can't do it, you'll have somebody hold you up.

The A-4 could not stand such a push. Erection takes place very slowly by utilizing two hydraulic lifts with four telescoped pistons. In spite of this you must make sure to stop erection at the proper moment so that the A-4 does not fall backward.

<u>For Erection You May Utilize the Rocket Carrier Only</u>

First you will have to remove the draw tongue and then relieve the front axle. Only then may you pull up the launching platform from the front. It serves as a footing for the erected A-4-

The A-4 is launched from an upright position.

REMOVAL OF THE DRAW TONGUE

(1) First turn up the triangular shutter located at the front axle. The steering of the front axle is now locked.

If you fail to do this,
the trailer will go sideways!

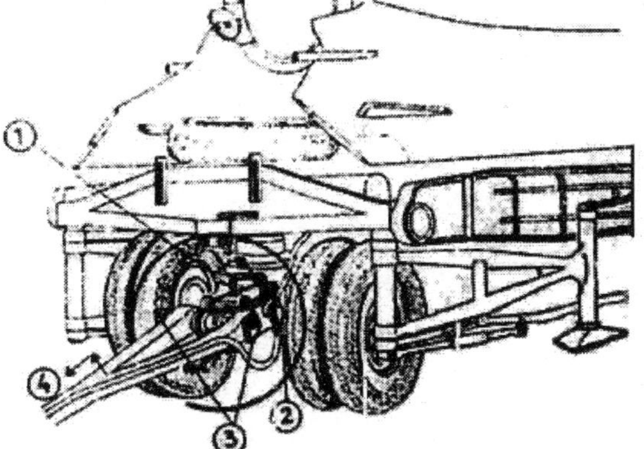

(2) Remove the air-brake hose and attach the dust covers.

(3) Remove the lock of the draw tongue bolt and the plate, lift up the fork in the front. Now it is possible for you to pull out the draw tongue bolt.

(4) Remove the fork-otherwise you will not be able to push the launching platform in front of the rocket carrier

Should you place the draw tongue on the ground then don't pinch the air brake hoses.

REMOVING THE WEIGHT FROM THE FRONT AXLE

(1) Loosen the bolts and turn the support-struts to the front.

(2) Fasten the struts on the carrier frame. You may arrive at the correct length of the struts by turning the coupling sockets.

(3) Turn your tie-spindles on the supporting plate until the inner doublex tires are free from the ground.

Otherwise—the front axle will not be sufficiently relieved of the weight.

Be careful that you have the right adjustment on the shift lever of the ratchet key.

Otherwise—the ratchet will turn in neutral.

(4) Put your axle on the machined strip and see if the front axle is level.

If you don't, the A-4 will never stand verticaly.

<u>Remember!</u> Always raise the lower side of the vehicle, do not lower the higher side!

Otherwise—you will apply weight to the front axle again.

The same holds true in Life:

Always raise your standard, never lower it.

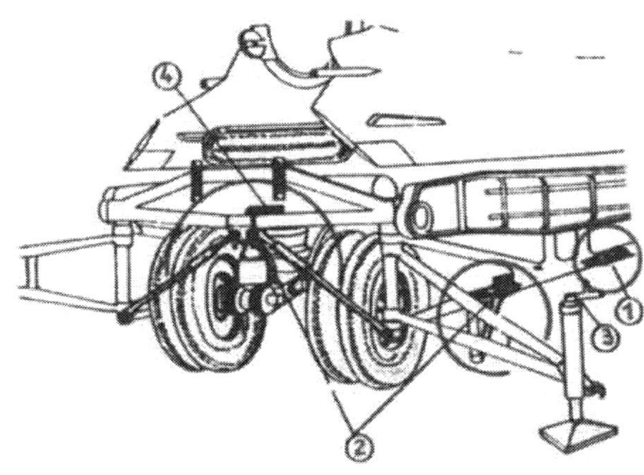

ERECTING THE LAUNCHING PLATFORM

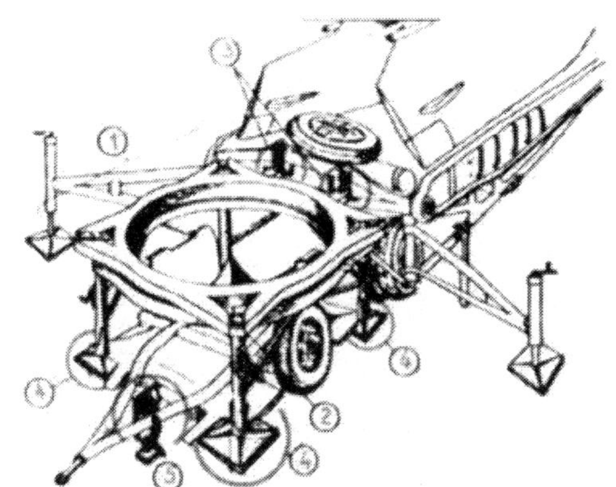

(1) Move up the launching platform. The main support assembly on the torret ring should be pointed toward the rocket carrier.

(2) The discs must be vertically above the legs. The terminal screws must be tight.

(3) Place the two belts into the bracket stands at the terret ring and into the guides at the rocket carrier.

(4) Unscrew the four foot plates of the launching platform.

(5) Pull out the jack handle and lower the jack until the chute is clean.

(6) Lower the table by raising the four foot plates.

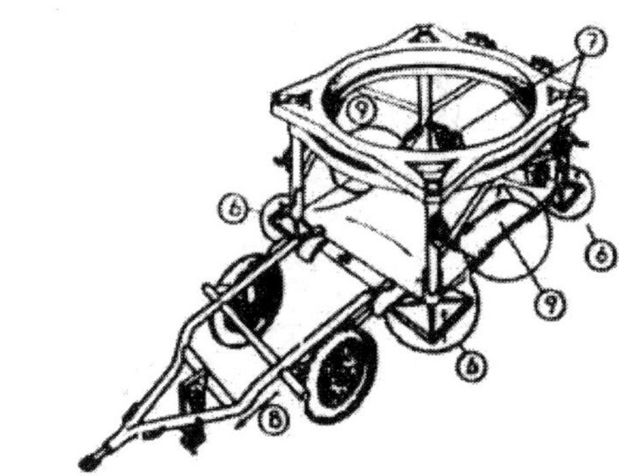

(6) Remove the two rear pins of the channels from the loop of the chute.

(8) Pull out the chassis and set it off to the side.

(9) Insert the wheel cut-out ratchet and place the launching platform into the lowest position.

<u>Otherwise</u>-Fin 1 and 2 will immediately bump against the discs should one unintentionally erect the missile.

Be sure that the weight is distributed on all four foot plates.

Before Transferring the A-4 onto the rocket carrier:

Check the oil-level in the oil tank of the tilting mechanism.

Before Erection:

Fasten the footbridges in the mountings of the tip frame.

Before Starting the Cold Meter:

(1) Turn cross-handle "left" to end position.

(2) Turn shut-pff valve "left" to end position.

(3) Turn change-over valve "left" to end position.

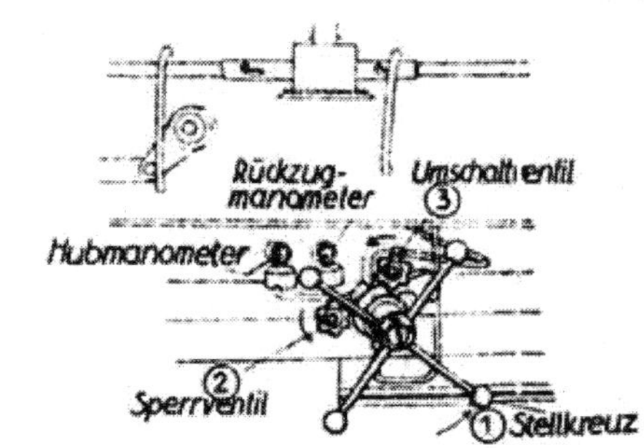

The Erection

In order to erect the "bird"
you must have hydraulic oil.
It is easily done with
atmospheric pressure.
For erection-the cross handle
must be in the left hand end position.

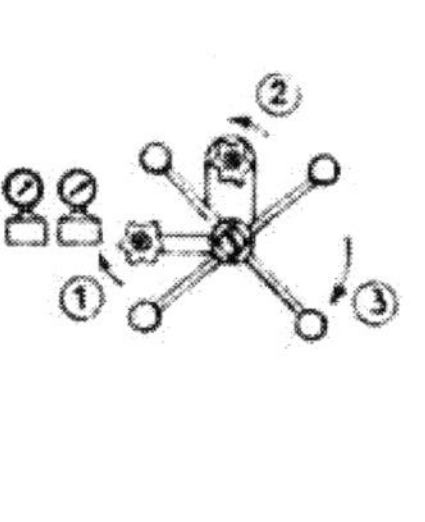

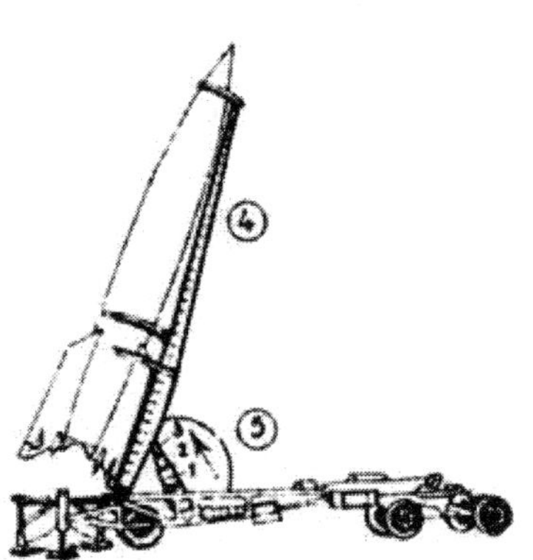

(1) Close the lock valve. Caution.

(2) Open switch valve. As soon as the meter runs.

(3) Turn your cross handle clockwise

(4) The tilt frame will then rise and
the bird will erect.

(5) Watch pistons 1, 2, 3--it's important.

(6) Now open the shut off valve.

(7) The retraction piston can come up now.

(8) Do this slowly, turn the cross handle to the left.

(9) Once the missile is in a vertical position you must turn the cross handle to the left end for stopping.

(10) Close the shut off valve again. You are through now.

(11) The tilt frame is now straight.

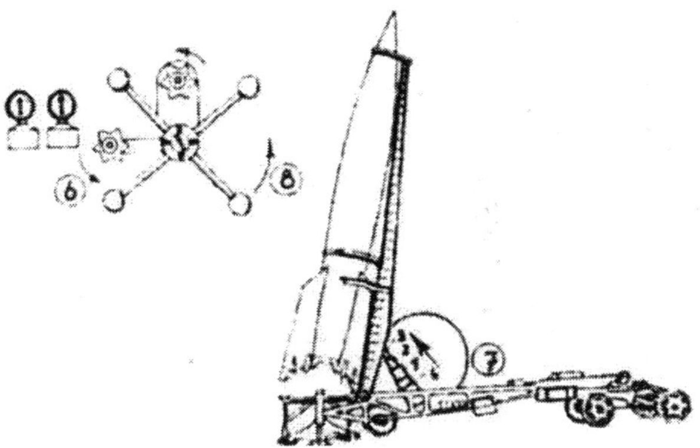

Note:

No one is allowed to be under the tip frame while erecting or lowering the missile.

Check the correct adjustment of the central valve, should it not be possible to fully erect the tip frame.

After erecting the missile, raise the table until the turntable is about 10 mm below the lower edge of the pin.

Check to see if the contact surface of the device rests properly on the discs of the launching platform and if the sockets at Fin #4 are open for the radio lead-in cord.

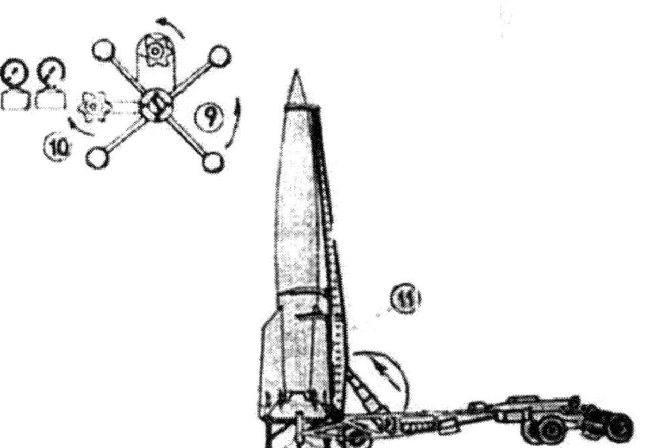

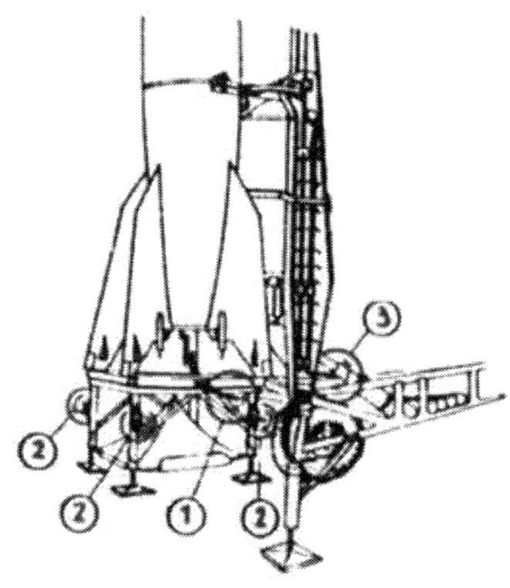

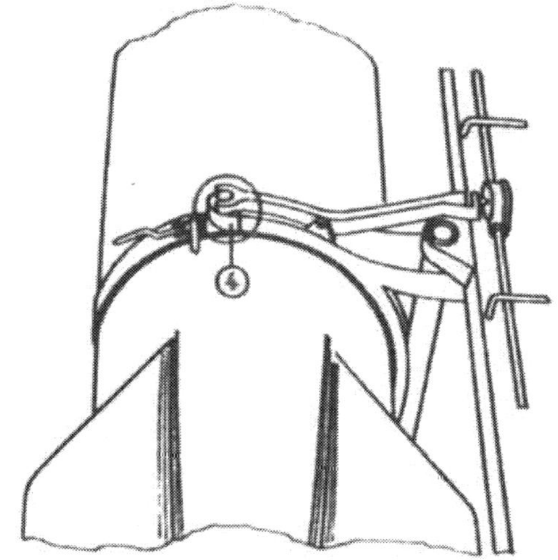

Remedy:

(1) Turn the turntable with the ratchets after loosening the assembling bolts at the mountings and removing the guides if necessary.
If necessary move the entire launching platform.

(2) Jack up the platform until it rests against the contact surface of the A.4. Lower the A-4.

(2) A ratchet is connected to the short propeller shaft and turned until the resistance decreases and the strap bolts are relieved.

(4) Open the strap from the footbridge and do not forget to remove the bracket joints from the connecting frame.

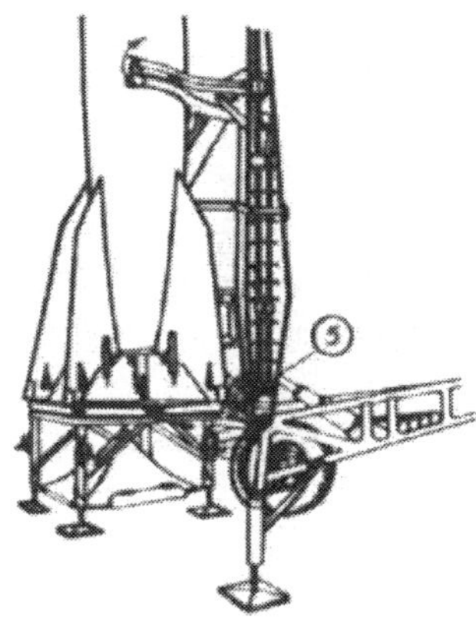

Remove the strap-belts from the connecting frame. The ratchet is attached to the long shaft and turned until the strap bolts are removed, from the connecting frame. Observe this procedure from the foot-bridge.

Should the A-4 be covered by a center tarpaulin, the strap belt will not hold it any more and may be removed now.

6) Open the clamp. A ratchet is attached to the square end of the worm gear and the clamp is opened. Pay heed that the clamp does not touch the device.

Otherwise—the tip-frame will rub against the device when being lowered.

__Remedy:__ Lifting or lowering the cantilevers of the rocket carrier.

PUSHING BACK THE ROCKET CARRIER 90 CM

Lower the tip frame approximately 10 degrees until the A-4 stands clear.

(1) Turn the change-over-valve clockwise to "stop"

(2) Turn the shut-Off valve counter clockwise to the left, the tilt frame will lower itself.

(3) After lowering up to approximatley 10 degrees, turn the cross handle all the way clockwise.

The tip frame will stop.

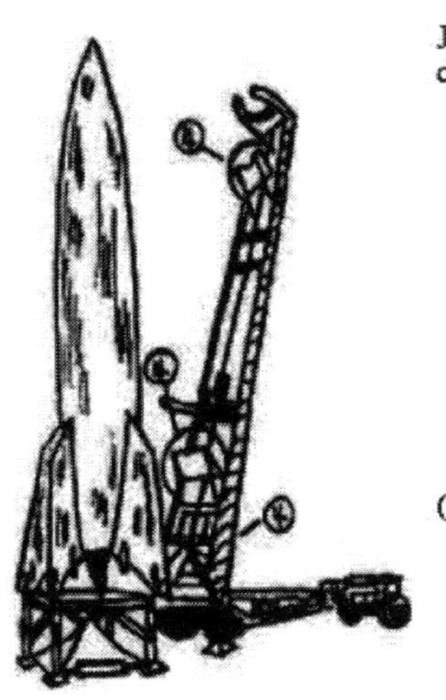

Jack up the support discs of the rocket carrier until they have cleared the ground.

Otherwise—you will not be able to push back the rocket carrier which is necessary for erection of the platform.

Push back the rocket carrier exactly 90 cm and pull the hand brake.

Give the A-4 a 90° turn to the left. Fin 2 and 3 will then point toward the rocket carrier.

(4) Turn the cable winch and lower the working scaffold. Relieve the front axle and level it by utilizing the water level.

Erect the tip frame until the edge of the upper working scaffold is spaced approximately 10 cm, from the instrument compartment. The valve settings are the same.

Shut off the meter.

Erect the chain rail.

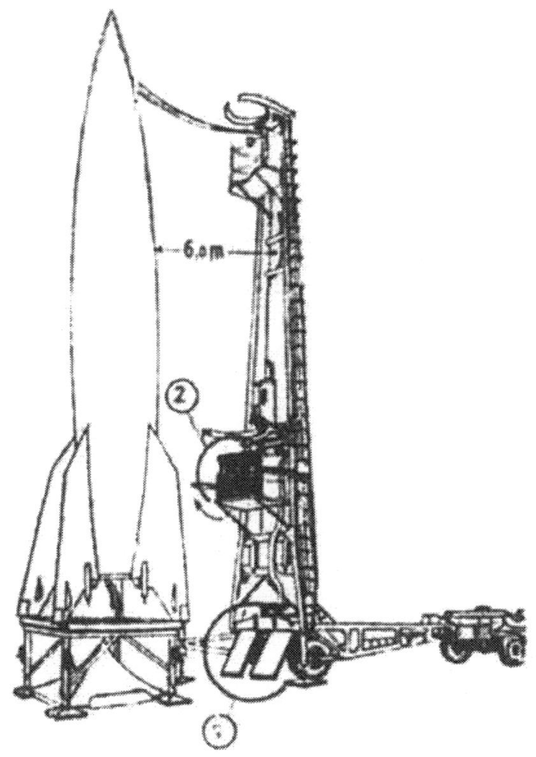

After fueling the A-4, that is just prior to launching, push back the rocket carrier approximatley 6 meters. The tip frame remains erected.

At the same time and according to the specifications given by the surveying crew, the A-4 is turned so that Fin No.1 will point in the direction of fire.

Don't forget to tighten the four terminal screws located at the turning ring of the launching platform.

(1) Place the guard-platform front of the tires of the front axle.

(2) Fold up the power lines of the magnetic plugs.

MORAL: The tip-frame mechanism places the rocket in vertical position without any difficulty

MOTTO: Often, one must go to the trouble of returning the unsuitable.

RETURNING

Should a defect appear on the erected A-4 which cannot be done away with at the launching site, then the fuel must be removed from the A-4 and the A-4 itself be placed on the rocket carrier. It is then transported to the field depot where the defect is removed.

(1) Push the rocket carrier toward the launching platform to within a distance of 15 mm. It is not necessary to insert the guides.

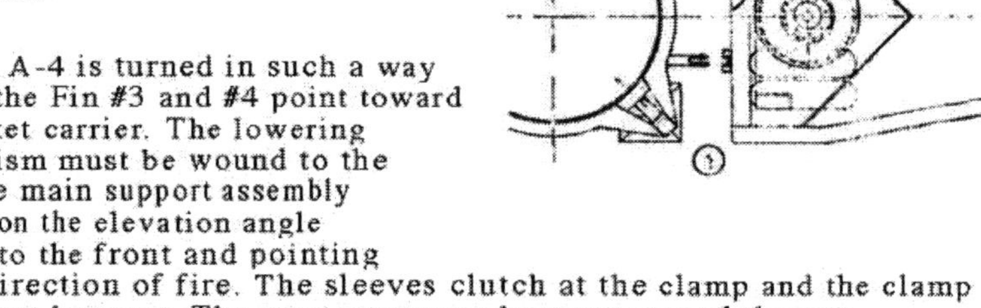

Watch that the A-4 is turned in such a way so that the Fin #3 and #4 point toward the rocket carrier. The lowering mechanism must be wound to the top. The main support assembly borders on the elevation angle located to the front and pointing in the direction of fire. The sleeves clutch at the clamp and the clamp itself must be open. The rope grummet plugs are wound the outside. The working scaffolds are fastened to the tip-frame. The upper part of the rope grummet is open. Move the tip-frame into a vertical position.

Slowly move into the final position and be careful not to drag the main support assembly or clamp on the side of the device when raising or lowering the crane of the rocket carrier.

Climb up the ladder and close the clamp.

You will raise or lower the launching platform by utilizing its four props until the rope grummet plugs and the connecting frame of the A-4 are even. This may also be accomplished by raising and lowering the crane of the rocket carrier.

Turn the rope grummet somewhat toward the A-4. Get up on the scaffold and see if the rope grummet pliugs can be inserted in the connecting frame.

Otherwise—you will damage the connecting frame by forced insertion of the rope grummet plugs.

Should the rope grummet plugs stick on the side, then turn the turning-ring of the launching platform.

Turn the long driving shaft until the rope grummet plugs are inserted in the connecting frame.

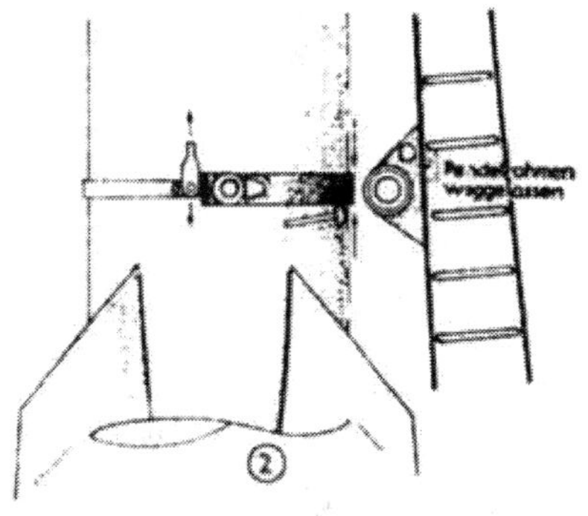

Hook the bracket joints into the connecting frame (2). Should the bracket be above or below the connecting frame then tap the lower part of the brace with a hammer.

Lock the upper part of the brace by moving the tightening lever and insert the locking pin.

Lower the launching platform approximately 4 cm.

Lower the tip-frame into the horizontal position.

Remove the working scaffoldings and fasten them together to the carrier frame.

Put on the tail tarpaulin

MORAL:

The things we don't like to do
must be done twice sometimes.

A soccer player has to observe the other playing members of his group so that he will turn the ball over to them at the right time.

Otherwise—victory will be questionable.

As a driving mechanism engineer you will have to observe the work of the other sections so that you are able to start your work and end it in conjunction with theirs, should the instruction chart require it.

Otherwise—there will be unnecessary delays.

MOTTO: If you want to win the ladies favor
you must think of ways to meet them.

All implements inside the A-4 make up the "en missile" installation.

All implements which stay on the ground after launching, make up the ground installation.

Erection of the ground installation as fat as the driving mechanism is concerned is done by you. Attaching the ground control box, the fivefold coupling, the oxygenreplenishing coupling and the ignition coil.

Connecting the ground installation with the installation aboard the missile must be done by you, for prior to launching you need the ground installation to operate the installation aboard the missile.

Help the trailer crew remove the draw tongue and move up the launching platform.

Remove the ground control box and five fold coupling from the implement trailer as well as your tools and hand lamps.

Start with attaching the ground control box as seen as the A-4 is in a standstill position. The extension tubes (1) of the launching position platform must be pulled out.

Prior to this loosen the screws completly (2)

Otherwise—you will not be able to pull out the extension tubes!

Tighten the screws as soon as you have pulled out the tubes.

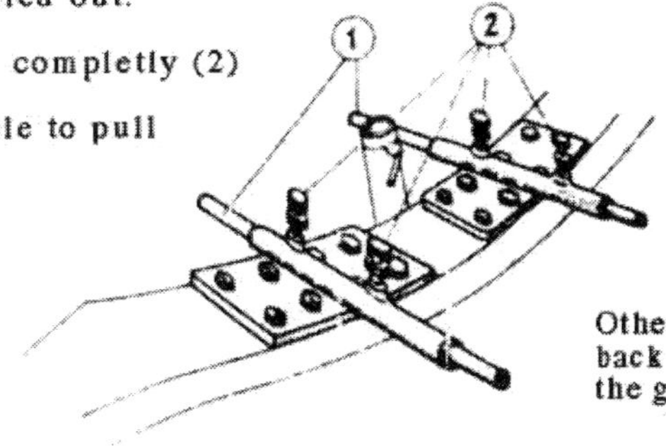

Otherwise--the tubes will push back while you are installing the ground control box.

Be sure that the door (1) of the box is facing out and the water tight plug (2) is pointing down.

Otherwise—the door won't open or the ground control box will be upside down.

Be careful that the extension tubes don't protude from the lug eyes of the lau or from those of the ground box,

Otherwise—the spacing is wrong, and the ground control box will t against the launching platform when turning the A-4-

Tighten the holding screws (3) on the plug eyes of the ground box.

Otherwise—the ground control box may fall off.

It is very important to correctly fasten the box at the launching platform.

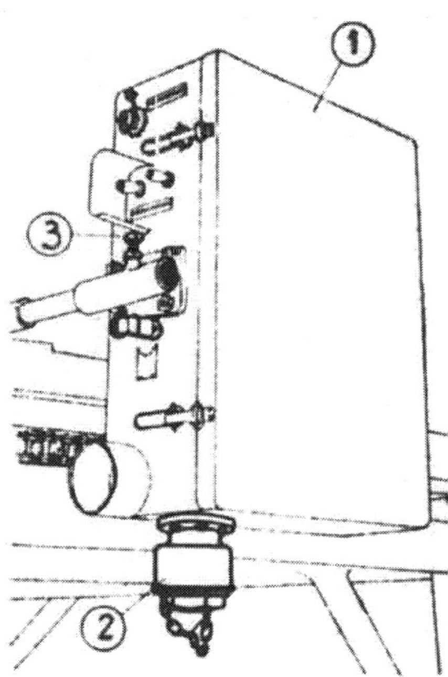

Install the fivefold coupling as soon as the A-4 is on the launching platform. Check to see if the rubber gaskets are serviceable and flush your lines with P-Substance. On the launching platform, between Fin #2 and #3 you will find the socket for the fivefold coupling. Remember 2, 3 and 5.

Loosen lever (1) prior to plugging in the base of the coupling device.

Loosen lever (2) on the device so that you will be able to push the coupling back and forth.

Loosen the coupling nut (3) at the universal joint of the counterpart so that the counterpart is turnable in every direction.

Turn hand wheel (4) to the left (as seen from the top), so that you can tighten it up later.

Pull back the protective cap (5) and fasten it with the locking pin.

Put the counter part over the five pipe connections of the exhaust firing of the A-4.

Don't force it.

Otherwise—the rubber seals will be damaged.

When inserting the counter part be careful not to tilt it.

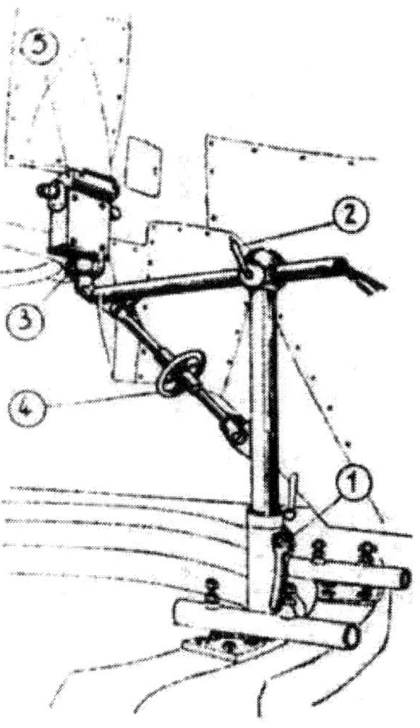

Hold the joining linkage together until coupling knot (3) and knob (1) and (2) have been tightened by your comrade.

Turn the hand wheel clockwise (as seen from the top) until the counter part is pressed against the exhaust fairing.

Ready the protective cap by removing the locking pin and checking it.

Otherwise-at take-off the rubber seals will burn.

Connect four of the five lines of the five fold coupling to the ground control box.

Now watch out!

1. The individual lines are marked and carry a tag with inscription.

2. The corresponding connections at the ground control box are also marked by tags.

3. The terminals for the regular main, the auxuliary control pressure line and the filling line are located at the back plate, theaeration connection at the right side of the ground control box.

Remove the safeguarding devices on the tubes and the dust cape on the terminals of the ground control box and check the sealing surfaces.

Screw the dust caps on the provided four bolts of the ground control box and carefully safeguard the protective coverings of the five lines.

Otherwise-they will get dirty or get lost.

Bend the tubes properly and fasten the lines to the designated point on the grou utilizing the lock nut. Move the tube back and forth slightly while tighte

Otherwise-the bevel will not fit and will leak.

The middle line will not be connected. B-Substance drips from it sometimes. Should B-Substance flow from it then report it to your squad leader.

Connect the compressed air battery of the FR Carrier with the high pressure connection on the left side of the ground control box.

Then report: Ground control box and fivefold coupling installed.

Should you forget to install the five fold coupling, then there will be no connection betwen ground and missile.

Inspection of the tail screwings: Tighten all screws and secure them.

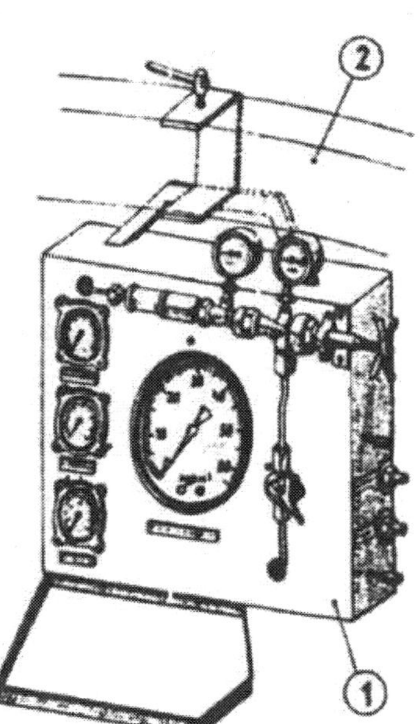

While your comrades are working on the ground, you will climb to the lower working platform. Take with you the following:

1. Your tools
2. Check panel, or board.
3. The connection lines for connecting the test manometer.

Hook your safety chains.

Otherwise—you will fall.

Fasten the check.board (1) on the cradle frame (2) of the rocket carrier. Open the four power unit shutters 1, 2, 3, and 5.

Store the screws well and don't let them drop.

Otherwise-- They will be missing when needed.

1) Loosen only the last screw. It must be a corner bolt.

2) Turn the shutter around this bolt and tighten the bolt slightly.

Otherwise-- The hatch may fall down on your comrad.

Report: Drive hatches open.

Load the compressed air battery as soon as you have received the order. Your comrad on the ground will open the hand operated closing valve on the P-Substance battery of the rocket trailer and that of the filling line on the ground control box (1).
Watch the slow increase of pressure on the high pressure manometer located below the T-Substance container.
At 200 atu call out: Out off supply.
Your comrad will shut the hand operated shut-off valve.

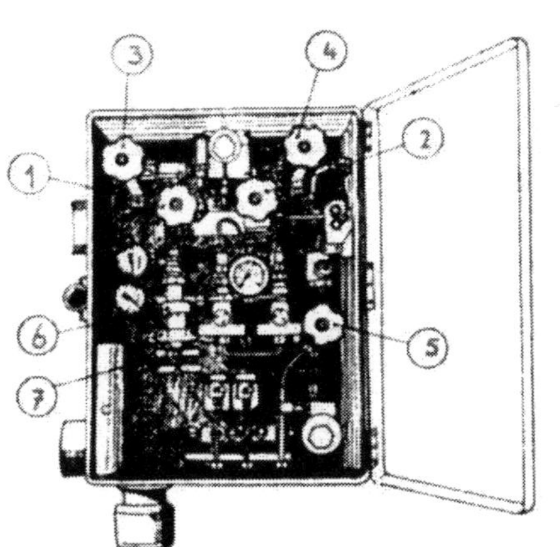

Your comrad adjusts the pressure reducing valve (6) located on the ground control box to approximately 23 atu by tightening or untightening adjusting screw (7). Listen to and feel the pressure lines. Should one of the screws leak then tighten it. Should you not be able to fix a leak then report it. Connect the control manometer (8) of the check panel to the provided for plug of the hand operated shut-off valve and open it. This valve is located at a junction of the low pressure line behind the pressure-reducing valve. Open the high-pressure hand operated shut-off valve located next to the T-Substance filler neck.
Set the pressure reducing valve of the A-4. Watch the increase of pressure at the low pressure manometer and tighten adjusting screw (3) at the pressure reducing valve where the lock screw (4) must be loose until the calculated value has been reached. The calculated value is taken from the accompanying papers. Listen to and feel the low pressure lines and remove leaks. At the same time make sure that the pressure in the compressed air battery is at least 180 atu.

Otherwise-the adjustment will be wrong.

Open the bleeder screw (1) on the front side of the safety valve by giving it a 1/4 turn to the left. The compresed air must come out of the bleeder screw.
Determine the amount from the control manometer.

Now there are three possibilities!

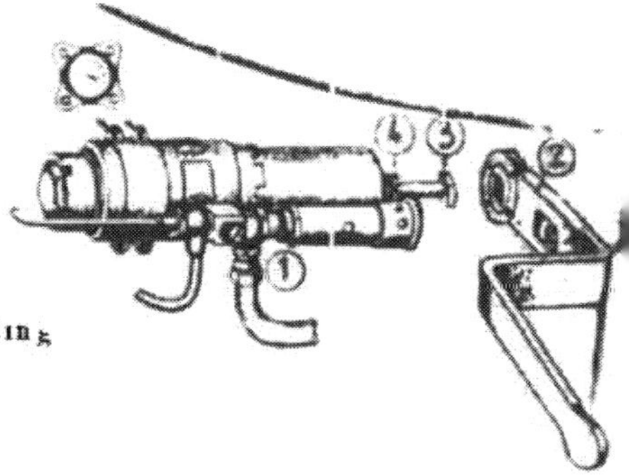

A. The determined value corresponds with the theoretical value to be used. Should this be the case then shut the bleeder screw and watch the pressure increase through the control manometer.
The thusly given value may not exceed the theoretical value by more than 1.0 atu.

Otherwise—the pressure reducing valve is not in order.

Loosen the bleeder screw again. The control manometer must give a reading equivalent to the theoretical value.

The difference may amount to 0.3 atu at the most.
Then the pressure reducing valve is in order.

B. The reading is under the theoretical value.
Should this be the case then place adjusting key (2) on adjusting jey (3), loosen the locking screw (4).
Turn the adjustment screw to the right and watch the increase in pressure through the control manometer up to the theoretical value.
Tighten the adjustment screw again.
Shut the bleeder screw and proceed exactly as described under A.

C. The reading is above the calculated value.
Should this be the case then place adjustment key (2) on adjustment key (3).
Loosen the locking screw (4).
Turn adjustment screw (3) with the key to the left and watch the reduction of pressure at the control manometer.
Let the pressure drop at least 0.5 atu under the theoretical value.
Now turn the adjustment screw to the right again and proceed as described under B and A.

MORAL: Flying or falling of the projectile depends on the job done here.

Checked are these who unite forever
MOTTO: You will check to find an error.

A doctor checks the heartbeat of a human being by using his ear. You check the operation of the valves in the propulsion unit by ear.

Propulsion Unit Test
With the propulsion unit test you test aeration, the ignition circle and circuit battery.

(1) Place your switchboard on the rotating ring of the launching platform next to the ground control box (2). Switch (7) "ignition circle" must be turned on.

(3) Put the plug into the counterpart located in the ground control box.

(4) Remove the safety cap from the connection of the aeration line of the igniter equipment.

(5) Open the hand valve of the auxiliary control pressure line in the ground control box.

Listen, if the A-Substance vent valve opens with a blunt knock.

Report: "Clear for Propulsion unit test!"

The launching leader will contact the launch control car by phone. He gives continuous orders.

The switch-sergeant handles the test-and firing switch located on the propulsion control desk.

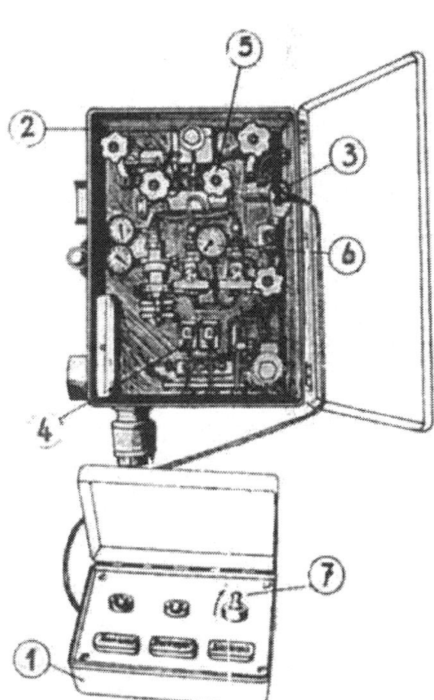

During testing time you will be lust eye and ear.

To the order: Launching valve on position #2:
Listen if the A-Substance vent valve closes with a blunt noise, and if the aerator valve opens and whether or not P-Substance enters the A-Substance container.
Watch the increasing A-tank pressure at manometer (6) in the ground control box.
The vent valve must close at 1.5 atu.

Should the A-tank pressure drop below 1.1 atu then the aerator valve must open

To the order: Listen if P-Substance is flowing from primer valve (4).
If it is, then turn off switch (7) "ignition circuit" located on the switchboard.

To the order: Launching valve on position #4:
Listen if the curcuit battery aerates with two rapid occurring hissing noises.

To the order: Turn off all switches:

Push back the switch marked "ignition circuit"

MAIN TEST RUN

(1) Place the lamp housing on the turning ring of the launching platform.

(2) Give the end of the cable comming from the lamp housing to your comrad on the lower working platform. He will in turn pull the plug of the steam generating plant and connect the cable.

Switch (4) "tank pressurizing" and (5) "pressure regulator" of the switchboard must be turned off. The ignition circuit must be turned on.

Again the launching platform leader gives his orders by phone to the launch control car. While testing, the following points must be manned.

one man at the lamp housing and switchboard,
one man watches the ground control box,
one man watches the pressure reducing valve (lower working platform)
one man on the upper working platform.

To the order: Launching valve on position #2:
Listen if the A-Substance aerator closes with a blunt noise.
See if the left lamp (3) "aerator" located at the lamp housing lights up.

At the switchboard the two first switches (4) "tankeration" and (5) "pressure regulator" are turned on and then turned off in the same sequence. When turning on the pressure regulator (5), the aerator valve must open. When turning off the tank-aerator valve (4)

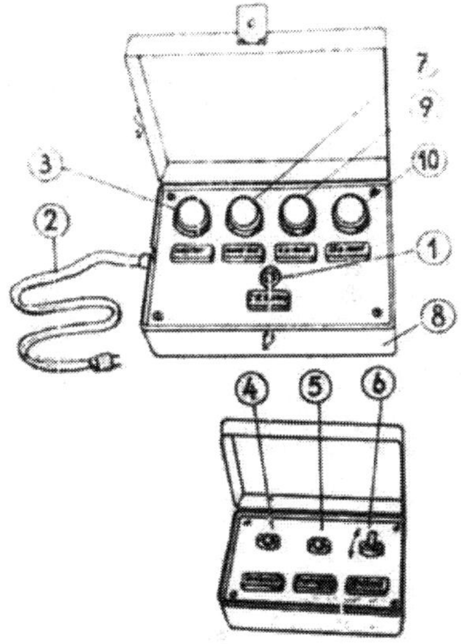

To the order: Launching valve on position #3:

Listen if the B-Substance pilot valve opens with a blunt noise.
Turn switch (6), "ignition curcuit", located on the switchboard, when the primer valve starts operating.

To the order: Launching valve on position #4:
Listen if the circuit battery will aerate with two short concecutive hissing sounds.

To the order: Launching valve on position #5:
Launching valve—check if lamp (7) "high pressure valve" lights up on the lamp housing. Then turn on switch (8) "Z-Substance contact" located on the lamp housing. Check if lamp (9) "8-ton valve" and lamp(10) "24-ton valve" light up on the lamp housing. Your comrad on the upper working platform will wait 40 seconds. Then the pressurizing valve must close with a blunt noise and activate the electromagnetic shut-off valve.

To the order: Combustion cut-off:
Check if lamp (10) "25-t valve" goes out.
Watch if lamp (9) "8 ton valve" and lamp (7) "high pressure valve" go out.
Should a lamp not go on or off, or should a valve fail to function then report this error immediatley.

To the order:
All Switches off: Disassemble the lamp housing and the switchboard. After the main test run there will be no parts interchanged in the A-4 or in the ground installation.
Otherwise: The main test run must be done over again.

MORAL: To test or be tested is not always the luckiest thing in the world.

Milk is the babies food
MOTTO: For fueling more experience is needed.

Once the main test run is completed, the A-4 is fueled.

Hereby you will work together with the members of the propellant battery.

Pay attention: Whatever is said about the four propellents of the A-4 on pages 104 to 106 concerns you too.

Be Careful When Handling Propellants.

To the order: Ready for Fueling

B-Substance: Climb to the upper working platform.
Screw the B-Substance fueling hose onto the nozzle of the B-Substance ascending line located at the radio trailer.
Remove the dust cap from B-Substance hose to the filler neck.

A-Substance: Put on Safety Clothes

Otherwise—your hands will freeze.

Climb on the lower working scaffold.
Connect the coupling of the A-Substance filling hose to the A-Substance ascending line of the rocket trailer.
Check the filling connections to see that they are clear of dirt, oil, or moisture.

Otherwise—you will have an explosion or the connection will freeze.

Remove the dust cap from the A-Substance filling nozzle.

Screw the filling plug connection with the large hand wheel (1) onto the filling valve.

Turn the small hand wheel (2) counter clockwise until you hit a stop.

Test the back pressure lids in the high pressure and emergency lines for tightness.

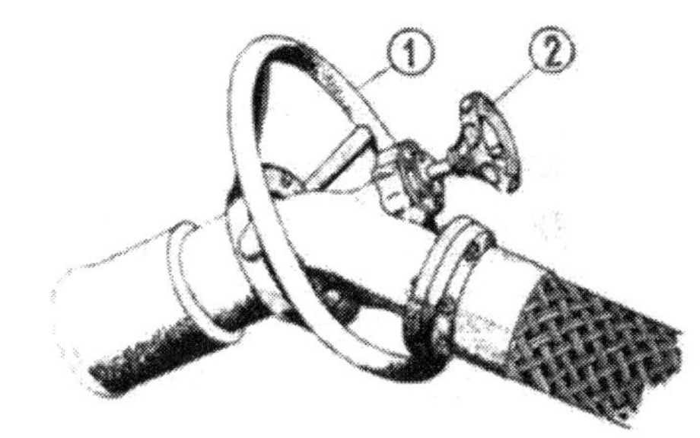

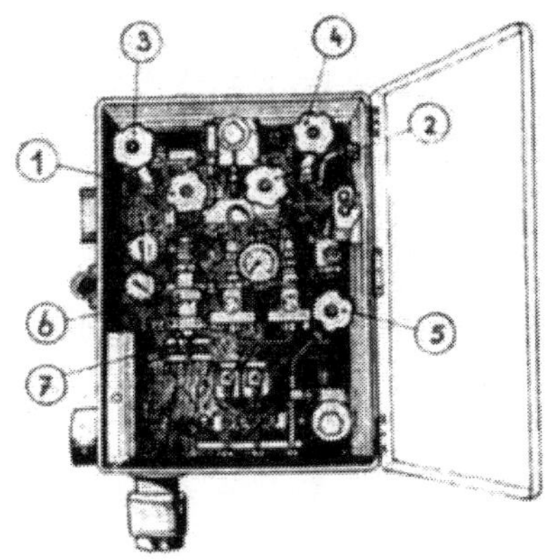

Shut supply valves (1) and (2).

Open bleeder valves (3) and (4).

Only the pressure remaining in the lines may now escape.

Otherwise-one of the automatic check valves is leaking and must be replaced.

Now close valves (3) and (4) again.

Open valves (1) and (2)

Otherwise—the A-Substance tank may burst.

Connect the extension pipe for the A-Substance bleeder tube and if possible dig a small drain ditch at the outlet.

Connect the pipe with attaching device.

Set up your hot air blower and connect the pressure hoses.

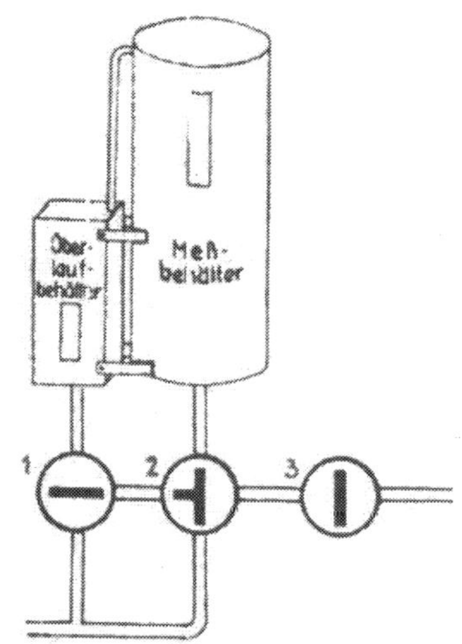

T-Substance: Have your water supply ready.

Put on T-Substance protective clothing.

Otherwise-you will be burned.

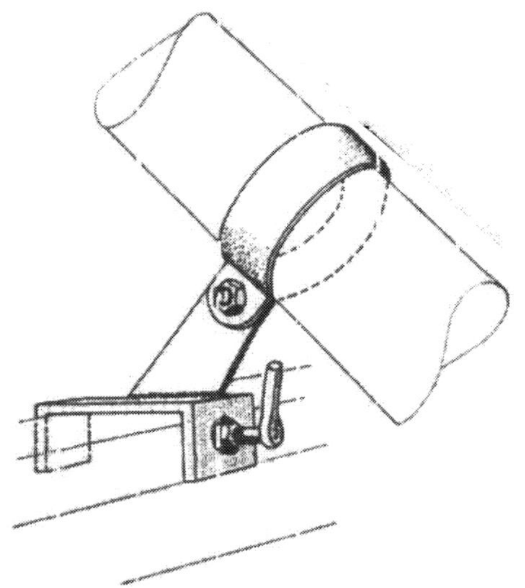

Remove the valve cone on the T-Substance filling nozzle.
Place the extension pipe on the filling nozzle.
Screw the one end of the T-Substance filling hose on the outlet nozzle of the measuring tank and the other end on the extension.
Adjust your valves located under the measuring tank as listed below:

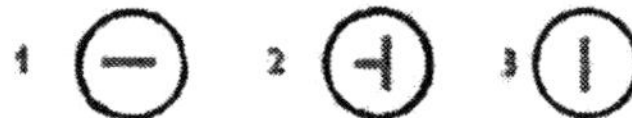

Report: "Ready for Fueling."

To the order: Fuel with B-Substance:

Check if the B-Substance lines are sealed. Keep the proper gaskets handy.

During fueling of B-Substance:

Fill the compressed air booster battery in compartment IV of the instrument compartment.

Climb to the upper working scaffold. Close the hand shut off valve located in the line between compressed air booster battery and B-Substance tank through the opening in the partition of compartment III.

Connect a line between the breather valve of the filling line of the ground control box and the filling sleeve of the compressed air booster battery. Install this line so that it runs up on the outside of the A-4. Your comrad on the ground will open the breather valve of the filling line in the ground control box. Watch the pressure increase in the compressed air booster battery through the manometer. At 200 atu give the order "shut off supply." Your comrad will shut off the breather valve of the filling line in the ground control box.
Disassemble the connection line between the breather valve of the filling line in the ground control box and the filling sleeve of the compressed air booster battery.

Open the hand shut off valvet marked fuel tank pressurizing at the B-Substance tank. Determine by touch whether or not the heating system of the pressure reducer is turned on.

To the order: Fuel with A-Substance:

Utilize the hot air blower and blow hot air at the serve unit shutters. Check if the A-Substance lines are leak proof. Manipulate the fittings with nothing more than a wooden hammer.

Otherwise—they will break off.

Move the small handwheel of the fueling fitting.

Otherwise—it will stick.

To the order: Fuel with T-Substance: Watch the increase in T-Substance through the eight glasses of the measuring and overflow container.

Otherwise-your comrad will not turn off the pump in time and due to over pressure T-Substance may squirt out.

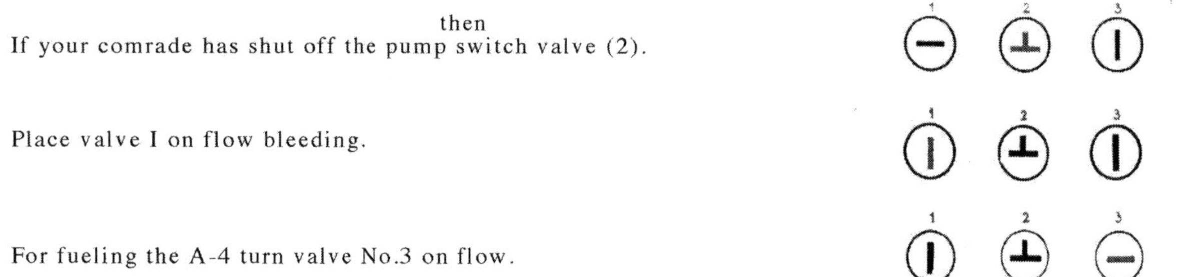

then

If your comrade has shut off the pump switch valve (2).

Place valve I on flow bleeding.

For fueling the A-4 turn valve No.3 on flow.

Observe the running over of T-Substance through the measuring glass within the filling hose.

If your turn valves 2, 1, and 3 No
T-Substance will be spilled.

<u>After Fueling:</u>

Remove the B-Substance filling hose. Turn the small hand wheel on the A-Substance filler neck coupling, clockwise until the pressure point has been reached.
Remove your A-Substance filling hose with the coupling.
Screw your safety lids on the A and B Substance filler neck nozzles.
Remove the T-Substance hose after you have drained it completely.
Lower your hose for rinsing. Watch that nobody is in the way.
Remove the extension piece from the T-Substance nozzle and tightly screw on the nut cap.

Otherwise—pressure will be lost here.

If you don't want to lose pressure
Secure your nozzle well.

The Combustion Unit

It was checked in the meantime by your platoon leader to see whether or not the welded seams and the soldered nozzles are leak proof and if the 18 cardboard coverings are spaced properly.

In addition he climbed the ladder and checked the lines located in the tail unit for leakage.

Otherwise—one may expect a tail unit explosion.

It is slightly difficult for a fat man
to squeeze into a combustion unit.

To the order: Fuel with Z-Substance:

Fetch the Z-Substance container from the pre-heating device. Be sure that the container has been heated well. Shake the ingredient well.
Remove the coupling piece from the Z-Substance filler neck.
Install the filling funnel.
Slowly fill up with Z-Substance.
Remove the funnel and lower it for washing.
Close the Z-Substance container with its coupling.

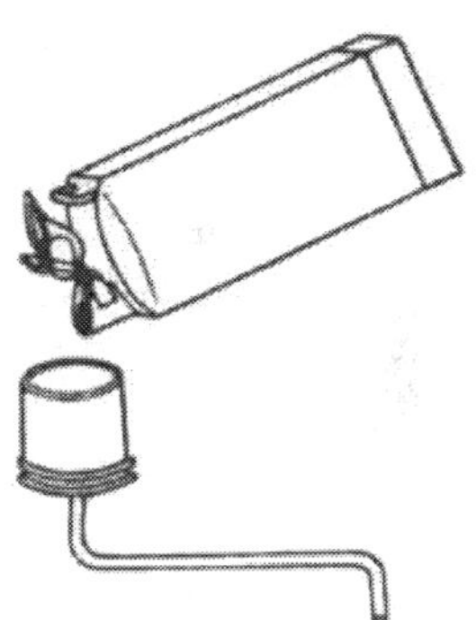

Otherwise--you will loose pressure.

Remember: Never fuel Z-Substance and T-Substance together!

Should Z- and T-Substance be mixed during fueling
Then you will soon be on fire.

Check the pressure of the P-Substance compressed air battery. It must amount to 200 atu.

Otherwise—you will endanger safe firing.

Check the theoretical value of the pressure reducer. Close the hand shut-off valve. Remove the control manometer.
Close all shutters. Be sure the shutters are seated properly and that no screws are missing.

Otherwise—the shutters will dismantle in air.

Anything which dismantles in air
was not checked prpperly before.

Install the A-Substance oxygen replenisher coupling.

Remove the valve cap on the A-Substance replenishing line between Fin #1 and #4. Loosen lock (1) in the groove of the launching platform.

Fit the base of the device into the groove.

Loosen Lock.(2).

Turn the hand wheel (3) to the right (as seen from the top)

Check the coupling and whether or not the sealing surface is undamaged and the coupling parts are free from dirt, oil, and dampness.

Align the counterpart to the replenishing line and press the counterpart against the upperpart of the coupling at the exhaust fairing.

Tighten levers (1) and (2).

Turn the handwheel to the left (as seen from above) until the counterpart rests snug against the upper part of the exhaust fairing.
Connect the compressed air piping from the pipe connection at the underside of the ground control box to the oxygen replenisher coupling.

Check for leaking lines when refueling.

Remove the extension pipe of the A-Substance breather tube when refueling has been completed.

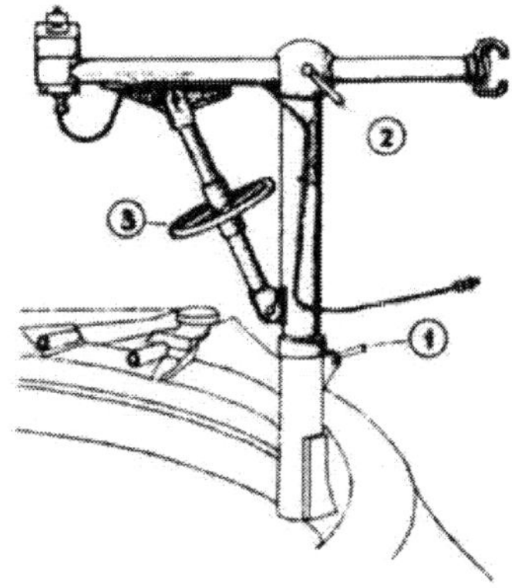

MORAL:

You confine dangers within their borders, if you are careful while fueling.

MOTTO: The ignition must wake the powers contained in the A- and B-Substance.

The energy contained in the A- and B-Substance is released only then when both propellants are properly ignited. Therefore you will need an ignition device.

INSTALLING THE IGNITER TORCH

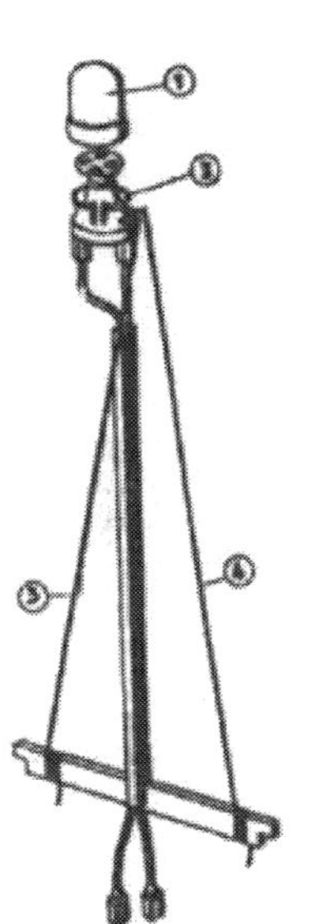

Unscrew dust cap (1) from the anvil head.

Don't forget to remove the plugs from the connection nuts of the connecting lines.

The connecting lines of the igniter torch will for once not be blown through.

Check magnesium band (2) for damage and tight connections. Check whether or not it is spaced evenly from the anvil head.

Check if the connecting wires (3) and (4) are kooked up tight and if they are spaced equally from the anvil head.

Remove the connecting wires and wrap them tightly around the crossbeam.

Otherwise—they will touch and ignition control is endangered.

Insert the igniter torch into the combustion chamber.

Fasten the horizontal strip in the tin-holders to the bottom side of the stern frame.

Be careful not to hit against anything.

Otherwise—the magnesium band will be torn off.

Connect the wires of the ignition control circuit to the clamps of the ground control box.

CONNECTING THE IGNITER FLASK

Put a bucket of water on the side.

Otherwise—you will not be able to help should T- or C-Substance be contacted.

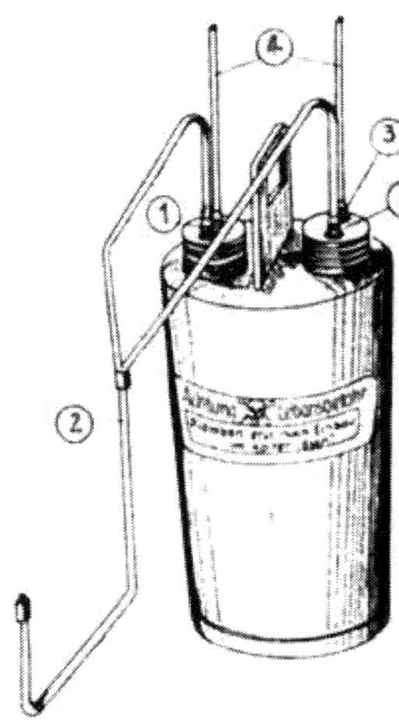

Put the igniter flask into the designated basket at the launching platform.

Remove the plugs from the breather connections (1) and screw in the pressure clamp (2).

Remove the connecting nipple from the drain nozzle.(3)

Attention! Always loosen the breech plug (1) first and then the female connecting nipple (3).

<u>Otherwise</u>— the ignition fluid may spray out because of overpressure.

Blow out the T- and Z-Substance feed line.

<u>Be careful</u>—the lines may contain some fluid residue left over from before.

Connect fuel line (4) to drain plug (3). Connect the fuel lines with the lines of the igniter torch.

<u>Only by Order!</u>

Connect the compressed air line from the ignition valve of the ground control box to line (2)

MORAL: Follow instructions well or your loved ones will mourn you.

THE ELECTRICIAN CREW

Electric batteries supply the airborne A-4 with current.

These batteries are installed at the launching site only.

Prior to take off of the A-4 the steering mechanism must function perfectly.

It must therefore be tested prior to take off.

Radio devices in the A-4 transmit signals during flight. They are given from the ground.

It is therefore necessary to test all devices prior to launching.

The electrician crew performs this job.

You know now just how much depends on your accuracy and conscientiousness.

MOTTO: A person who catches on slowly usually suffers from a short circuit of the brain.

CURRENT STROM

Current is needed in order to operate the radio and steering devices of the A-4.

Current gives the A-4 the order for take-off.

Current makes it possible to send orders to the airborne A-4.

Current is needed by the A-4 in order to follow orders.

Current ignites the fuse.

Current ignites the high explosive upon impact.

Your source of the current in the A-4 is:

1. the 27-volt course gyro battery;
2. the 50-volt course gyro battery.

You are to install them in the rected A-4.

In order to preserve the force of currents in the A-4 there are several sources of current installed in vehicles. They deliver the current as long as the A-4 is on the launching site. Under utilization of special cables and plugs you will deliver the current to the A-4. The two plugs are called magnetic plug #1 and 2.

The moment of take-off is especially important.

It will show you if you have done your job correctly.

Important devices in the A-4 and on the ground must start to operate at this moment. They start to operate due to take-off contacts that signal the takeoff of the A-4. One contact signals the A-4, the other the ground installation.

The instrument compartment harbors several delicate devices. Protect them from dust and dampness when you open the doors to the instrument compartment.

PREPARING THE MAGNETIC PLUGS

Prior to the erection of the A-4 remove the magnetic plugs from the transport frame of the Rocket Carrier.

Otherwise—you are forced to pull the connecting plugs up to the working platform alongside the erected A-4.

Be careful that upon erection of the tip frame the cables to the magnetic plugs do not get caught in the frame of the F R Carrier.

Otherwise—the cable will break.

TAKEOFF CONTACTS

Before you erect the A-4 fasten the counterpart of the missile mounted takeoff contacts on the launching platform under fin 2 and then fetch the ground takeoff contact.

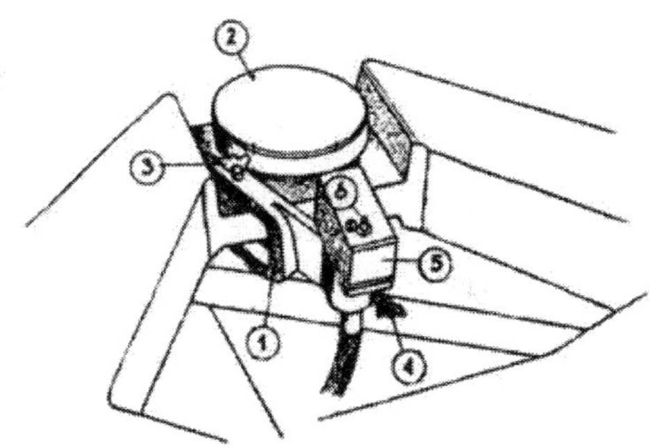

Upon erection of the A-4 push its clamp (1) located adjecent to plate (2) under fin (3) from the inside of the launching platform.

Tighten screw (3).

Untighten screw number (4), push switch (5) from below, against the fin in such a way that the push button (6) is depressed.

Then tighten screw (4) again.

Connect the cable of the takeoff contact to the socket to the ground control box.

INSTALLING THE FIFTY-VOLT COURSE GYROBATTERY

At low outside temperatures store the battery in the test car so that it will not get cold. Take the battery and check for a 50.4-volt potential. Don't forget to press the inductance button on the volt meter from 10 to 20 seconds.
Open door III to the flight control compartment of the erected A-4 and pull up the battery with the elevator. Install the battery and tighten the mounting stand well.
Make sure the connections between battery and bus are correct + to + - to -.

Otherwise—you are forced to reverse poles and waste time.

INSTALLING THE TWO 27-VOLT COMPRESSED AIR BATTERIES

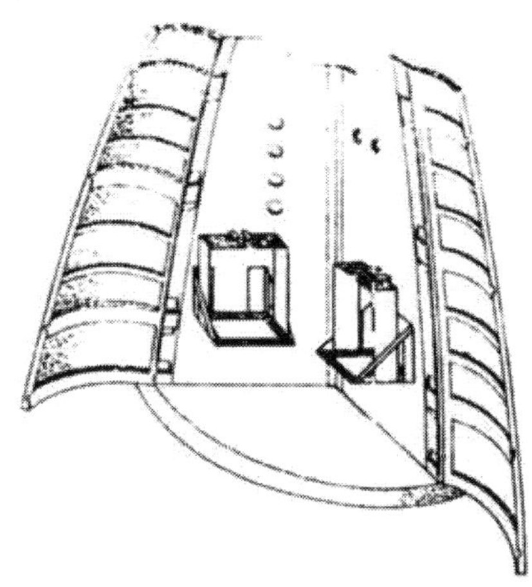

At low outside temperatures, and before installing, these batteries are to be placed in the test car. Fetch them by utilization of the volt-meter test them to make sure they are fully loaded.

The load of each battery must then contain 16 volts. Do not forget to press the inductance button of the volt meter 10 to 20 seconds.

Test to see that the poles are not grounded.

Open door I to the instrument compartment of the A-4 and pull up the batteries.

Both ends of the battery are to be connected from plus to minus. The missile mounted terminal boards however are to be connected + to + and - to -. Otherwise—you are forced to reverse the poles later on and, cause delay in launching.

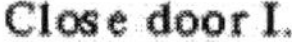

Close door I.

INSULATION RESISTANCE

At the counterpart of the magnetic plug 2 measure the insulation resistance between the Plus-Terminal Board against earth as well as the Minus-Terminal Board against earth. The resistance must be about 500,000 ohms. Your meter is sensitive. Handle it with care.

PLACEMENT OF THE MAGNETIC PLUGS

Grease the housing of the magnetic plugs. Plug in magnetic plug 1 to section 2 of the instrument compartment first. Then plug in magnetic plug 2.

Otherwise-if you don't they will not hold.

Should the plugs fall off, the current necessary for the holding magnet will be missing. Check the ground source in the power supply carrier. Until launching time keep the plugs from falling off by fastening them to the upper working platform by means of the chain.

MAIN TEST DRIVE

During this test the two magnetic plugs will fall. You must prevent them from falling down completely. Climb to the upper working platform and loosen the safety chain. Hold the magnetic plug cables approximately one meter from the plug and pull slightly in the direction of the plug. Hold on tightly when the magnetic plug falls out.

After the test do not forget to plug in the magnetic plug again. In addition you are to check and see that the batteries under stress do contain the prescribed amount of voltage (27 or, as the case may be, 50 volts.)
Moral: During work never forget enemy, heat, dust or dampness.

When it is too late people often say
MOTTO: I wish I would have steered a little better.

THE STEERING DEVICE

Shooting itself is not hard but to hit the bull's eye is. The barrel directs the rifleman's bullet and so does the gun barrel, the shell. The A-4 is not shot from a barrel and therefore needs a different kind of guidance. A steering device is used for this purpose.

Your automobile would run off the road should you not steer it correctly, or should the steering device be defective, or set wrong. The A-4 has automatic flight control. If you have set it correctly it will operate automatically upon takeoff.

The A-4 has 8 Rudders, 4 of them lie in the air stream, there are air rudders, 4 lie in the jet of liquid fire, caused by the departing flue gas, called jet-rudders. The air rudders are thin and are located at the outer lowest edge of the four fins. The jet rudders consist of graphite and are mounted below the combustion unit.

Ziel

1

2 4

3

Sometimes the air and jet rudders operate together, sometimes separately.

The four fins are numbered: Fin 1 points in the direction of the enemy. Should you be behind the launching platform, and the installed A-4, that is behind Fin 3 then remember:

Fin 2 is to your left,
Fin 4 to your right.

The air rudders 2 and 4 are called trim tabs. You cannot adjust them by hand. They are to even out constructive errors.

MORAL: Especially when launching a Rocket a missing steering device will ring disaster

MOTTO: With the pitch gyro and roll and yaw gyro the course of the A-4 is plotted.

THE PILOT OF THE A-4

A pilot boards a boat. He knows exactly what course to take, the location of the sandbanks as well as the direction and strength of the current. He safely guides the boat past all dangers to the harbor and point of destination. But in order to steer correctly, the pilot must know the location of the boat. Two pilots are aboard the A-4.

They accompany the missile from time of launching to combustion cutoff. Then they have completed their job. They give the commands to the individual rudders of the steering device. They know exactly when the A-4 deviates from the prescribed course. They are called pitch gyro and roll and yaw gyro.

<u>The pitch gyro</u> is installed in a level casing. You'll find it in section III of the instrument compartment. You can get at it only by way of the upper working platform. The pilot eye is located in this casing and reacts toward any deflection from the prescribed course. It is a fast moving gyro fastened to two rings in such a way as to enable movement in every direction. Such a gyro has the characteristic of remaining in the original position. Should an unusual deflection occur between the gyro and its housing, the gyro will notice it.

<u>The roll and yaw gyro,</u> the second pilot of the A-4, is housed in a vertical casing. It, too, may be located in section III of the instrument compartment and also contains a movable gyro. It prevents the A-4 from turning longitudinally and prevents a side deflection from the desired path of the trajectory. Pitch gyro and roll and yaw gyro must be placed in a certain position on the A-4. The housing may therefore be adjusted by turning the adjusting screws

If the gyro notice a deflection from the desired path of the trajectory of the A-4, an electric current, or so-called control current, is released. This affects the prime mover of the rudders (vane or servo motor) in such a way that the A-4 is brought back. Within this circuit there is a control amplifier which, so to speak, gives its commands to the individual rudders according to the reports received from the two pilots.

Pay attention to the following:

The gyro are able to perform their job only if they are seated correctly. Upon starting up it takes 3 minutes until the gyro have seated correctly.

Remember it takes you sometimes several minutes to get situated.

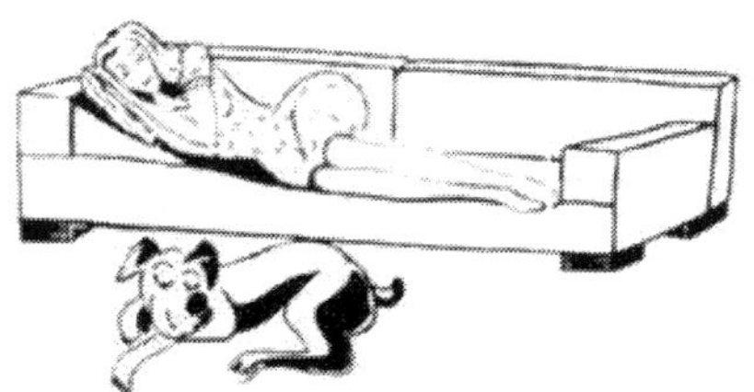

MORAL: Even we must often admit nbt being capable of stating our course or whereabouts.

MOTTO: At last it shall be made clear,
What to do to make it clear.

It takes a soldier 20 minutes to ready his sailboat and prepare it for sailing. The trip through wind and waves may then begin.

To clear the steering device you have 20 minutes time. If you finish all your duties in the prescribed time you will save time.

While clearing the steering device, test interferences artificially through air gusts etc., just as hey occur during flight.

During testing of the steering device the A-4 must be motionless.

Contact the launch control car by phone. Here the corporal in charge of the controls watches the angular displacement of the control surface and also the control current at the control desk. He will give you the orders.

Removing the Fin Shoes
Prior to erection of the A-4 loosen the screws and remove the fin shoes.

Installing the jet rudders
Set up the 4 rudders so that they are within reach.

Check the rudders closely. Cracks, irregularities and loose frames make them useless.

Screw on the rudders.

Do not forget to place the 4 spacing washers at the end of each rudder. Tighten the screws well. Install the safety plates.

Do not fix the rudders now but ready rudders 1, 2, and 3 for fixing.

Insert the jet rudder bracket (No.1) from the inside of the launching platform into the provided for connecting pipe (No.2) so that the rudder nose (No.3) is located in the groove of the mounting stand. Do not use force. Should the bracket not fit, then loosen the screws (No.4) and readjust the retaining clamp.
Now remove the jet rudder brackets and place them under the respective jet rudders.

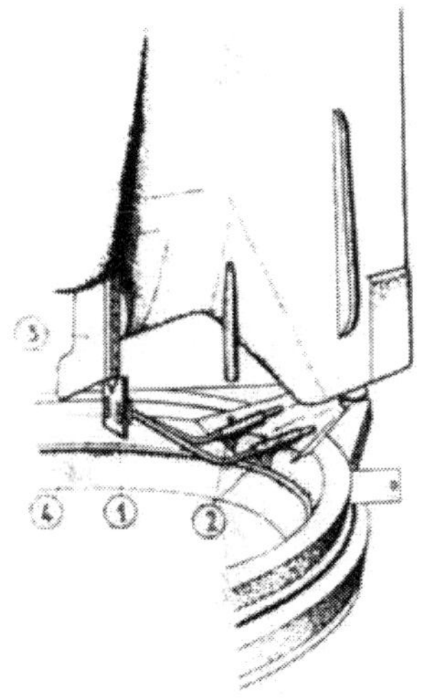

Adjusting the Steering Device.

To the order: Rectify control current No.1, you open the shutters to the four vane meters.
Change the potentiometer in the vane alignment box until the order "stop" is given.
Do likewise with regards to control current #2, 3, and 4.

To the order: Level.
Loosen the locknut on the leveling bracket, turn the two adjusting screws uniformly until the order "stop" is given.

To the order: Adjust roll and yaw gyro. Loosen the locknut on the roll and yaw gyro bracket, turn the adjustment screw until the order "stop" is given.

To the order: Rectify control current #1 and #3.
Do as you did above.
You must adjust the steering device by turning and testing as little as possible.

<u>Synchronization and testing of the vane meters</u>

After receiving the order to do so, check if both rudders move equally fast towards Fin #3 and then towards Fin #1 even when pressing against them. Then report by telephone-synchronization clear. Check if the rudders #1 and #3 move about the same from O to Fin #2 and then to Fin #4. When pressing against them pay attention to the sound of the operating vane meter. Precision adjustment for synchronization of the collaborating vane meters and for the standstill in the zero position is next. Watch the synchronism of the rudder pairs and make corections by adjusting the balance potentiometer. In doing this you must produce a standstill of the rudders within the zero position without interference.

Testing the Set-up

As soon as given the order adjust rudder #2 and #4 according to Fin #3. Watch the path of the rudder. The "switch" corporal in the launch control car counts from 21 to 24. After that rudder #2 and #4 must move towards Fin #1 evenly and then stop even when pushing against them.

Testing the Trim Steering

When given the order adjust rudder #1 and #3 somewhat contra.

Then check if the trim tabs move in the same direction that rudder #1 and #3 are standing.
Then report: "Trim tabs move correctly".
If the trim tabs return to the zero position immediately, then report : "Trim steering clear".

General Test Run

After receiving the order: "Take off", (this is approximately 2 seconds after the magnetic plugs have fallen off) remove the take-off switch and count aloud from 21 to 24.
Check if rudder #2 and #4 run out to Fin #1.
Then report: "Program clear".
Turn rudder #1 around the zero position and watch the direction of the trim tabs.
Then report: "Trim steering clear".
Place the take-off switch into the old position. Tighten the screw well.

After Fueling

Upon fueling and renewd erection and after turning the A-4 in the direction of fire, the control currents must again be rectified if necessary.

Take-Off

For take-off jet controls #1, #2 and #3 must be fixed.

MORAL: Once the last one has finally caught on
Then instead of "clear" it will soon be "naturally".

FREQUENCY DOUBLER AND COMBUSTION CUTOFF

MOTTO: Dear Friend this A-4
Is really great,
Because you can get rid of the
"spirit" at combustion cutoff.

If the "spirit", meaning the "B" rocket propellant, is ignited, then it will burn together with the "A" rocket propellent and propel the A-4. Acceleration will increase steadily. The propulsion unit will be cut off at a certain speed. In order to accomplish this you must induce a combustion cutoff. The fire in the combustion unit will go out and the A-4 will continue on in the same way a shell does after leaving the gun barrel.
The higher the velocity at combustion cutoff, the further the A-4 will fly.
The firing range may be changed by cutting off combustion at various accelerations.

For giving combustion cut-off, via radio, the combustion cut-off flight instruments are:

No.1: Frequency Doppler

It receives a certain number of electrical frequencies per second and sends back twice as much. The frequency doppler is needed in order to determine the velocity of the A-4

It is simiiat to the clerk in the bank who accepts a ten mark bill and then in return hands out twice as many five mark bills.

No.2: The cutoff Signal Receiver

It must stop the driving mechanism and this at a certain combustion cutoff. The receiver will transmit the sound only at a certain sequence of radio waves.

You are to test and watch over the two flight instruments.
The test may start as soon as the magnetic plugs are plugged in. Prior To this you must however perform certain preparatory duties.

Duties to be performed in advance

On the still horizontal A-4 check and see that the correct flight instruments are installed according to the frequency. Open shutter #1 of the instrument compartment.
Should, for excample, the order of frequency be "green 3", "red 6", "301", "Anton" then test the following:

No.1: Frequency Doppler
Check if the frequency doppler No.3 is installed and see that the connecting nut for the antenna binding posts are tight.

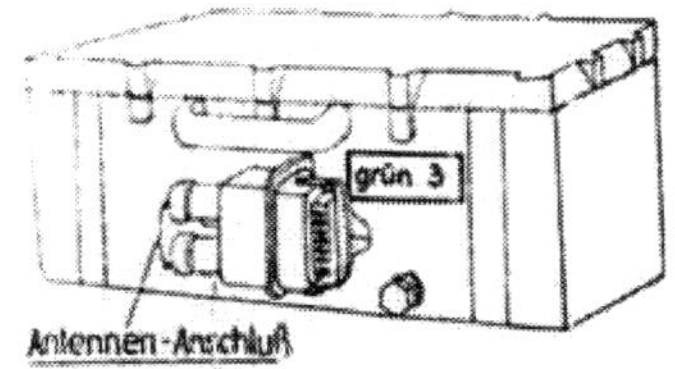

Remember: Should the frequency doppler be connected with another number, then change the entire device. It is not necessary for you to readjust the doorframe antenna.

No.2: Cut-off Signal Receiver
a) Check if a cut-off signal receiver with the label of group number 301 is installed.

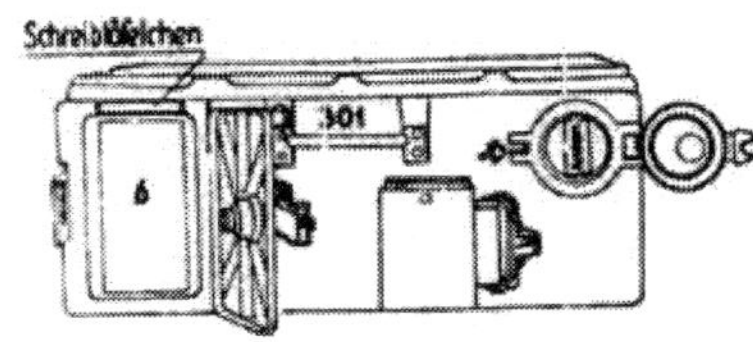

Remember: If another number is marked on the receiver then change the entire device.

b) Check if a high frequency charge #6 is installed in the cut-off signal receiver (left rectangular shutter)

Remember: Should a high frequency charger with a different number have been installed, then the receiver itself is not changed. You will however change the high frequency charger. Shut the shutter to the high frequency charger well.

Otherwise: the cut-off signal receiver will be useless during flight.

c) Check if a sequence selecter plug "Anton" has been plugged in. (right round shutter)

Remember: Should, for excample, a sequence selecter plug "konrad" have been used, then interchange it with "Anton". Again close the shutters well.

c) Check that the trim condensers on Fin 2 and 4 are set correctly.

At a frequency of red 1---3 set on mark 2
red 4---6 set on mark 5
red 7---9 set on mark 8

Therefore, after receiving the frequency order "red 6", set the trim condenser on 5.

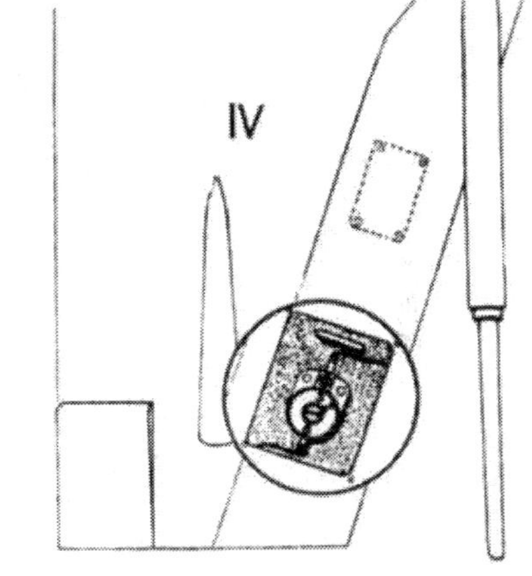

To the order: "Ready test car for flight instrument testing".
Move the test car 30 to 50 meters away from the A-4 and gead it into the direction of fire

PAYING OUT AND TESTING THE CABLES

(1) Place the power line cable 220 V/50 Hz from the test car to the power supply trailer, connect it to the B-Frame in the test car.

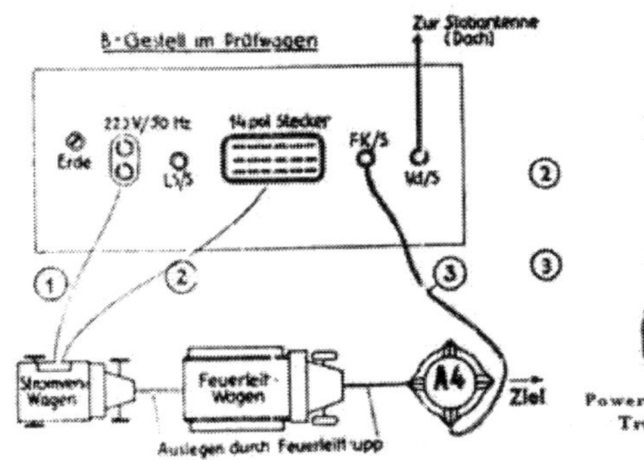

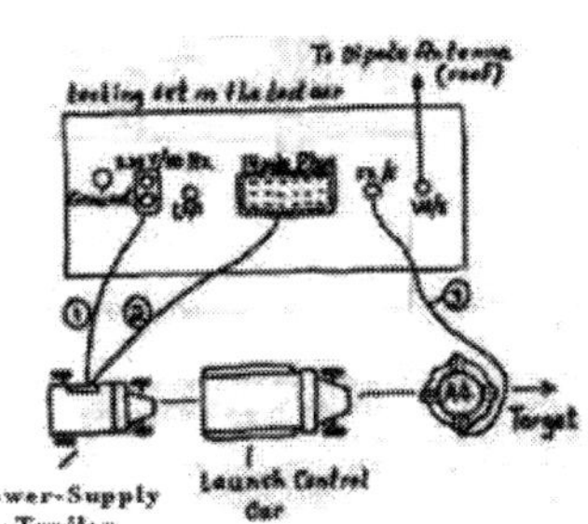

(2) Place the connecting cable (signal line) from the test car to the power supply trailer (relay box) and connect it to the test car. It leads to the 14 pole plug.

(3) Place the high frequency cable from the test car to the A-4 (Fin 4).
There you will protect the cable from burning by utilizing asbestos.
Do not connect the high frequency cable to the apparatus yet. The cable should not be bend. By utilization of the Isolavi-instrument measure the isolation resistance. It must be larger than 10,000 ohms.

At a lower measurement value the high voltage cable is to be changed.
Immidiately after erection of the A-4 plug the high voltage cable (plug) into the right jack on Fin #4 and protect the antenna plug from being torn off by utilizing the cable holder on the launching platform.
In the test car, measure the grind resistance with the Pantavi Instrument. It is to be smaller than 10 ohm.
At a correct value, connect the high voltage cable in the test car to the high frequency linkage "FK" at the top-side of the "B" structure.

<u>Remember:</u> Should the value be larger than 10 ohm, then measure the signal antenna of the A-4 to Fin #4 first and this by utilization of the Pantavi Instrument. The resistance must be between 0.4 to 1.7 ohm. By doing this you will be able to determine if the defect is in the connecting cable or in the A-4 or as the case may be in the cut-off signal receiver. Insert the dipole antenna into the provided for holders located on the roof of the test car.

<u>The "B-Frame"</u>

The "B-Frame" in the test car is utilized for testing the doppler and the cutoff signal receiver. It contains numerous fields and looks like this:

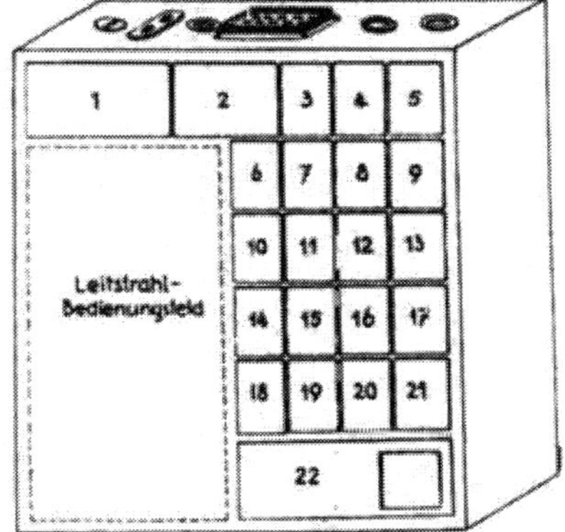

Field #1-net supply and monitoring field
" #2-frequency change stage
" #3-field of operation transmitter "red"
" #4-field of operation transmitter "green"
" #5 is empty
" #6 to #13-transmitter "red" for alternating frequency
" #14 to #21-transmitter "green" for alternating frequency
" #21-cutoff signal modulator

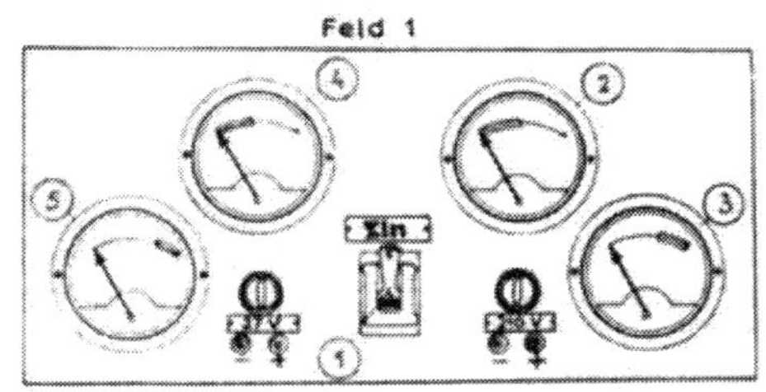

PUTTING INTO OPERATION

Field No.1-Place switch #1 on "in"

See that the deflection of the pointers of the measuring devices (2) and (3) for circuit pressure are within the "red" zone.

<u>Remember:</u> At deviation, be sure to request the proper current from the current supply trailer.
The deflection of the pointers of the measuring device (4) and (5) for DC voltage and current must lie in the "red" zone.

<u>Synchronization of the Vo-measuring device</u>

After being told what frequency to use, plug the test oscillator "green 3" into field #4. Check if the test oscillator operates in Field #4.

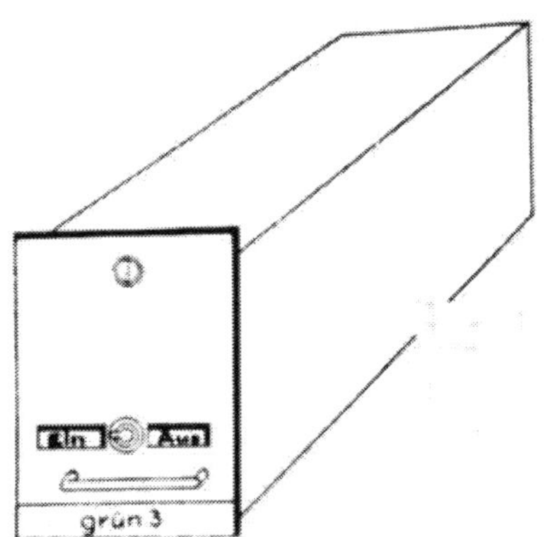

1-Turn the rocker-arm of the test oscillator in Field #4 on "red".
2-Take the doppler test case to the roof of the test car and fasten the fittings to the foot of the antenna rod in such a way that the measuring device will produce a maximum deflection. The deflection must lie within the blue zone.

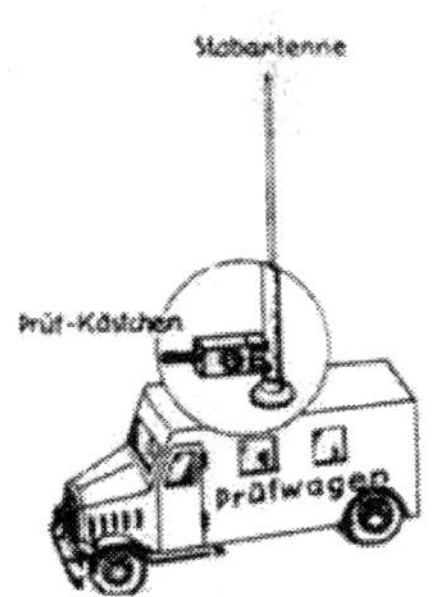

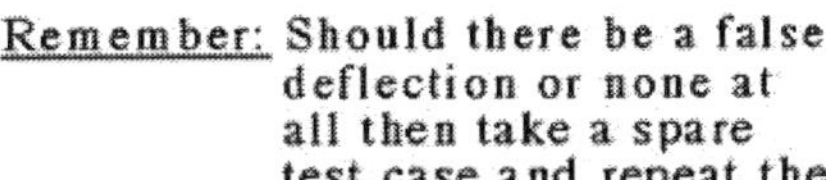

<u>Remember:</u> Should there be a false deflection or none at all then take a spare test case and repeat the test. Should you again obtain no results, then the error will lie in the doppler test oscillator.

3. Disconnect the test oscillator in Field #4.

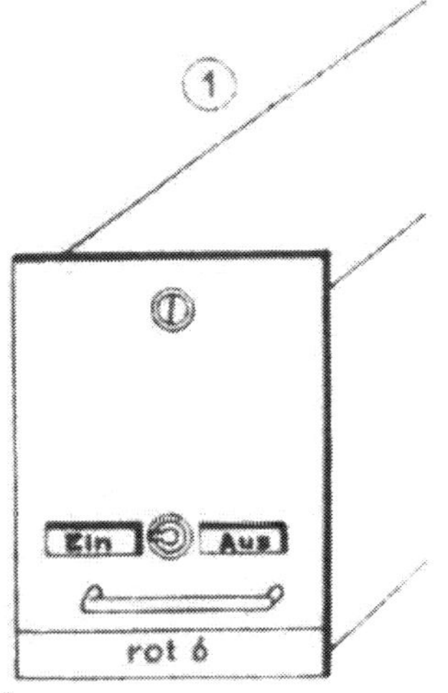

SYNCHRONIZATION OF THE COMBUSTION CUT-OFF DEVICE

(1) After being informed of the frequency plug in test oscillator "red 6" into Field 3.
(2) Open the shutter in Field 22.
(3) Push the lever up in Field 22 so that it points to "in"
(4) In Field 22 connect the series plug 301 with the cutoff signal modulator after receiving the correct frequency.
(5) In Field 22 connect the designated sequence selector plug "Anton" with the series plug 301.

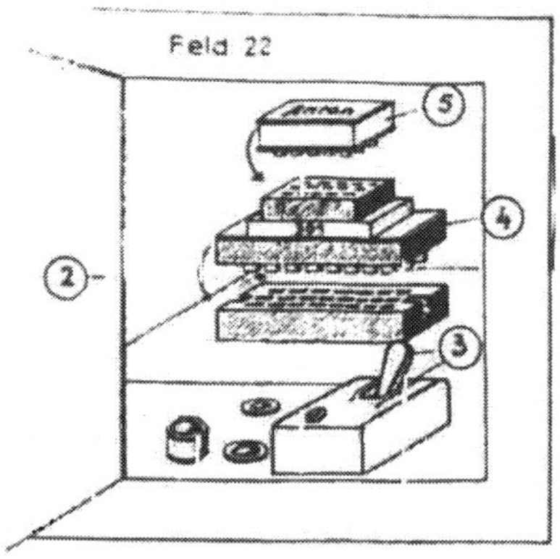

CHECKING THE HIGH FREQUENCY

In Field 3 place the lever of the test oscillator on "in".
In Field 2 place lever (1) on "in" and lever (2) on the high voltage marker.
In Field 2 check if the deflection of the pointer of the measuring device (3) is **within the red zone.**

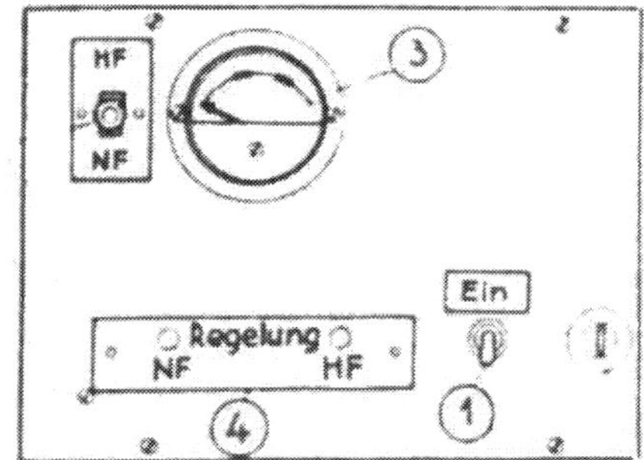

Remember: Should the deflection not be correct then regulate the high frequency adjustment (4) with a screw driver.

CHECKING THE AUDIO FREQUENCY

In Field 2 place lever (2) on low frequency. Check if the deflection of the pointer of the measuring device (3) is **now in the "green" zone**.

Remember: Should the deflection not be correct then regulate the low frequency adjustment (4) with a screw driver.

In Field 2 place the switch lever (2) on the middle position.
In Field 3 place the switch lever of the test oscillator on "off".

TESTING THE DOPPLER

Remember: The test may start as soon as the magnetic plugs are connected. Call the launch control car and give the order: "Connect the Hydrant", (Hydrant is a code for a type of combustion cutoff system for guided missile V-2). Due to this, Transformers (3) will start up in the A-4. Doppler and cutoff signal receiver receive current.

Climb to the upper working platform of the launching-car with the Doppler test case. From there you will call the test car and order: "Connect the test oscilator green".

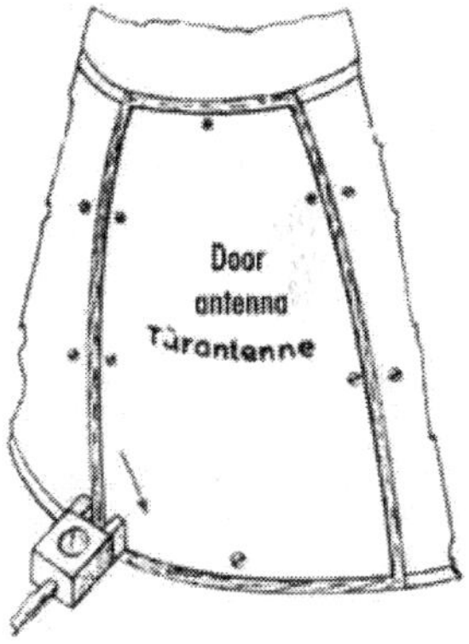

Check the door frame antenna with the Doppler test case.
The deflection of the pointer must lie within the blue region.

Remember: Should the deflection of the pointer of the test oscilator Not return to its original position, after being disconnected, then the Doppler must be inter-changed.

Door
antenna
Türantenne

TESTING THE CUTOFF SIGNAL RECEIVES

Testing may begin as soon as the magnetic plugs are connected. The actual test takes place in the launch control car. After receiving the order you will connect the following switches, located on the B-frame

1. Switch located in field number III (Transmitter red)
2. Switch located in field number II (Frequency changer)
3. Switch located in field number 22 (Out-off signal modulator)

a) Unarmed test
Check if the deflection of the pointer on the B-frame in field number 2 goes through the green region when the switch is in a downward position.

b) Armed test
The B-frame necessitates no checking
After completion of the test disconnect all switches as soon as you are given the order to do so.

MAIN TEST RUN

Five minutes prior to test begin plugging in all instruments.
Check if after removal of all magnetic plugs in field number 1 of the "B-frame" the deflection of the pointers of the measurung device (2) and (3) return to zero.
After completion of the test disconnect all voltage from the B-frame.

LAUNCHING

After receiving the order, hook up the B-frame again.
The B-frame is to be put into operation at launching time for the following reasons:
It is not possible for the magnetic plugs to fall and the ignition to be on, the A4 however, due to an error not take off.
By radio you are then able to immediately give combustion cutoffs by way of the high voltage cable.

MORAL: Without radio and radio communication a combustion cutoff at the proper time would be impossible.

MOTTO: It is important for you to know that by utilization of the J-Device combustion cutoff devices are quite unnecessary.

Years ago one was forced to look up at the tower clock to know what time it was.

If the tower clock was not visible and if you could not hear it one did not know just how late it was. In the days gone by however, this was not very important. Today one utilizes a pocket watch, this is much more convenient, one is not therefore dependent upon the tower clock. Our combustion cutoff system corresponds with the tower clock. We do not however measure the time but the velocity of the airborne A4, for we must cut off the power of the A4. For we must cutoff the power of the A4 at a certain velocity in order to make the A4 reach the required firing range.

The enemy will now try through his own radio devices to jam our radio communication between cutoff ground and aircraft apparatus. A velocity testing device was therefore sought (which compares with a pocket watch) mentioned above and which operating completely independent is carried by the A4. This device is not affected by any interference from the outside. The apparatus is called "integrater" of "J-Device". At a set cutoff speed it cuts out the driving mechanism of the A4.

Remember: If the J-Device is utilized no combustion cutoff apparatus is installed, and it is not necessary to set up the cutoff ground installation.

When installing the device, JG 1/1-3 you are to perform the following duties.

1. Adjust the support of the J-Device and bring it into the correct position.
2. Hang up the J-Device and connect it.
3. Engage the J-Device.
4. Adjust and test the cutoff speed (firing range)

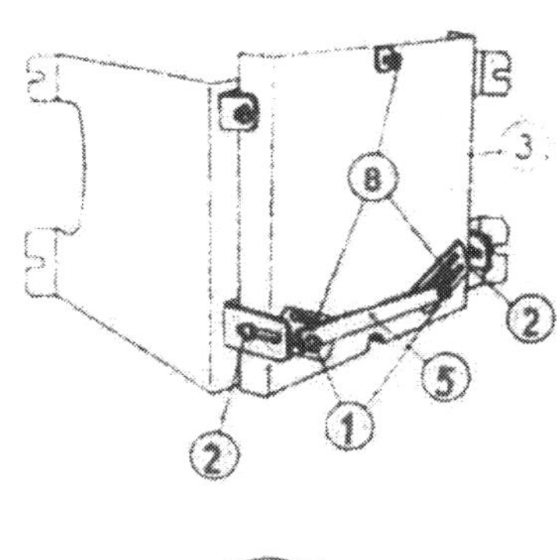

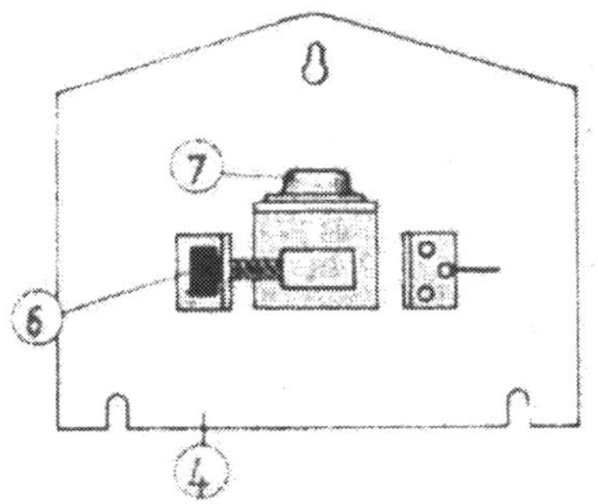

is used for adjusting the support of the J-Device 1/1--3.
You will use the clinometer for setting.
Loosen the retainer screw (1) of the supporters rotatable machined strip.
Loosen the swing bolts (2) of the rotating part of the frame (3).
Fit the adjustment device (4) into the support.
Be sure that the lower edge of the adjustment device is seated upon the rotatable machined strip (5).
Fasten the adjustment device by tightening the three retainer screws (8) so that they may be easily turned by utilization of the precision adjustment drive (6) for tilting to the left, or as the case may be to the right.
Turn the precision adjustment drive until the airbubble in the clinometer (7) is in the center.
Tighten the three retainer screws (8) of the adjustment device and bolt screws (1) well. At the same time check if the airbubble of the clinometer (7) remains in the center. Turn the rotatable support (3) in such a way that the air bubble of the clinometer is exactly in the center of the red control circle. The air bubble should not more than touch the edge of the control circle.

Otherwise-the adjustment device is not set accurately enough.
Tighten the swing bolts (2) of the adjustment device well.
Recheck the position of the air bubble in the clinometer.
Should the air bubble still be located in the center then loosen the three retainer scrws (8) and remove the adjustment device from the support.
You have hereby adjusted the support.

INSTALLING AND CONNECTING THE J-DEVICE 1/1--3

Install the J-Device.
Make sure the lower edge rests upon the rotatable machined strip.
Tighten the three retainer screws.
Plug the 20-pole plugs into their respective sockets.

ENGAGING THE J-DEVICE 1/1-3 FROM THE LAUNCH CONTROL CAR

Engaging the JG 1/1—3 is a prerequisite for adjusting the cut-off speed and testing the adjustment.

For this purpose you will utilize the integrator testing device calibrator 1, known as JS 1. It is installed in the center part of the radio controlled desk (FT desk) of the launch control car. You may connect the J-Device only when the magnetic plugs are inserted in the A-4 and when the launch control car is being supplied with current.

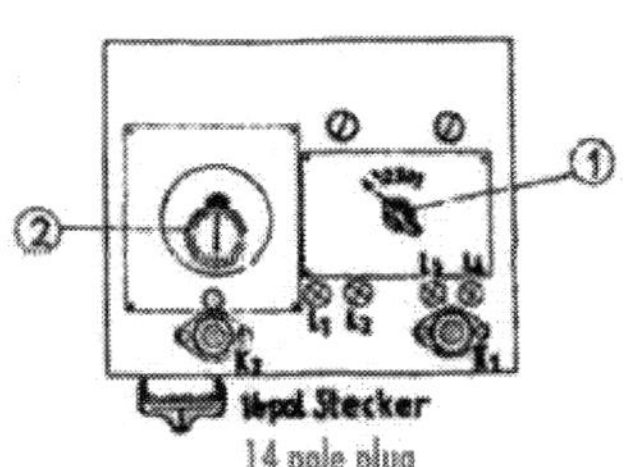

14 pole plug

Turn the rotary switch (1) to position O. The control lamp L 1 must light up after the JG 1/1—3 has operated for 40 seconds. You have now turned on the current for the integrator testing device calibrator.

The missile mounted JG 1/1-3 is supplied with current by transformer 3. It is connected by turning the left rotary switch on the radio control desk.

Wind the stop watch (2) and when it runs place it in the compartment with the glass window of the integrator testing device.

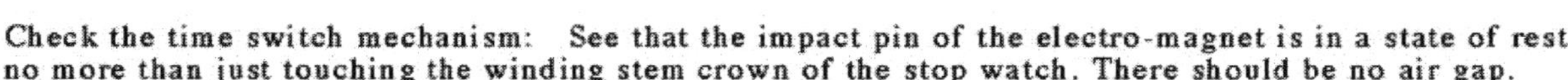
Check the time switch mechanism: See that the impact pin of the electro-magnet is in a state of rest no more than just touching the winding stem crown of the stop watch. There should be no air gap.

<u>Otherwise:</u> the test will not be accurate enough.

Place switch (1) for controlling the stop watch on position 1.
Depress the button K 2.

The impact pin of the magnet hereby pushes against the winding stem crown of the stop watch causing it to stop.

Again depress the button K 2.
The hand of the stop watch hereby jumps to O.
Place switch (1) back on the preparatory position O.
Wait until the all-clear signal light L 2 shows the test readiness of the JG 1/1—3.

ADJUSTING AND TESTING THE CUT-OFF SPEED

You may start testing only after the all-clear signal lights L1 and L2light up.

In the launch control car: Turn the switch (1) of the integrator testing device calibrator on position 1.

In the A-4: Set the range card value of the cut-off speed on the rough-and fine pointed dial of the JG 1/1-3.

In the launch control car depress button K 1 until the all-clear signal light L2 dies out.

This marks the test beginning of JG 1/1-3.

The light L3 (preliminary cut-off signal) and L4 (main cut-off signal) light up at an interval of approximately eight seconds.

The stop-watch is stopped when light L4 lights up. Read off the running time of the stop-watch and compare it with the time given on the range card.

In the A-4: Reset the five pointed dial according to the specification on the range card, should the two time periods differ.

You are then required to test the J-Device again but wait until the all-clear lamp L2 lights up again.

See that the two lamps L3 and L4 are properly timed.

Should it be necessary adjust the tine pointed dial again and make a third test.

Remember: The J-Device is ready for operation only if the all-clear lamps L1 and L2 are lighted.

CONCLUSION

In the launch control car place rotater switch (1) of the integrator testing device on position S.

MORAL: You have completed testing the device only After the two timing elements correspond.

The artillery soldier follows the path of the girl.
MOTTO: The A-4 follows the path of the Giude Beam.

GUIDEBEAM INSTALLATION ABOARD THE AIRCRAFT

The hunting dog follows the tracks made by game. He does not turn to the right or the left of the path.

The A-4 follows the path of the Guide Beam. It does not turn to the right or left of this path.

An electric path is produced by the Guide Beam ground device. It is called the Guide Beam plane.

It leads to the objective by way of the launching site.

The Guide Beam installation aboard the aircraft determines whether the A-4 is to the right or left of the Guide Beam plane.

It causes a deflection which turns the A-4 back on the Guide Beam path.

The flight of the A-4 is therefore much more accurate and the deviation at point of impact much less. The Guide Beam has no bearing on the firing range.

<u>Very Important:</u> The Guide Beam method is utilized at only part of the launchings, and you are to ready the device for operation in those cases only.

It is your job to test the Guide Beam ground device without operating its transmitter.

<u>Otherwise:</u> The enemy could get your bearings.

You will therefore utilize a model transmitter located in the B-frame of the test car.

DUTIES TO BE PREPARED IN ADVANCE

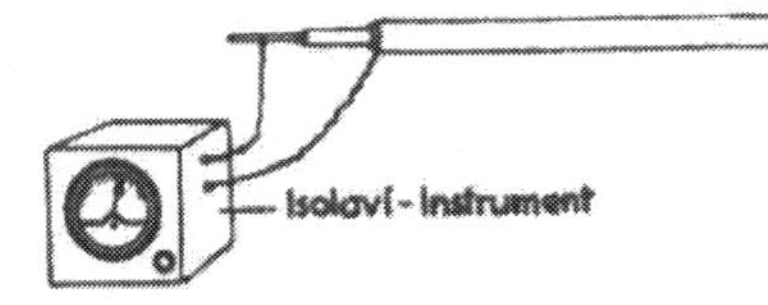

Check if the test-car is parked approximately 30 to 50 meters from the A-4. Check if the high voltage cable 220 V/50 Hz is laid out from the test car ro the current supply trailer and connected to the B-Frame. Check if the high frequency cable is laid out from the test car to the A-4. The cable should not be connected or bent.

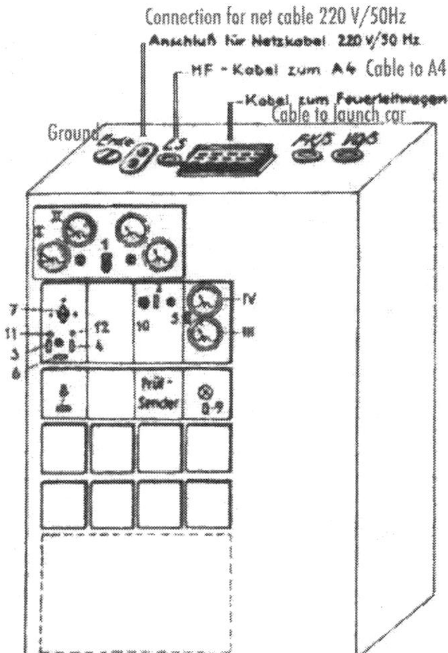

By utilizing the Isolavi-Instrument check the insulation value of the HF-(high frequency) cable.

It must be higher than 10,000 Ohms.

Remember: At a lower value changes the frequency cble.

Upon erection of the A-4 fasten the antenna rods to Fin 2 and 4.

Connect the HF-cable to the antenna rod of the Fin 2 or 4 and connect the clamp to earth.

Measure the drag resistance in the test car with the Pontavi-Instrument.

It must be smaller than 10 Ohms.

At correct value connect the high frequency cable (H-F) in the test car to the H-F connection "L S" at the upper part of the B-Frame.

OPERATING THE B-FRAME

The B-Frame in the test car is also utilized for testing the Guide Beam ground device.

Here adjustments are to be made in the left fields only.

See that switch (8) is in the middle position and switch (6) is to the right.

Place switch (1) in an upward position. You have hereby turned on 220 Volts.

Check if the pointers of the volt meters 1 and 2 are within the red zone.
Remember: At *a* fluctuation request the proper voltage from the current supply trailer.

OPERATING THE MODEL TRANSMITTERS:

Check if the HF-Transmitter is installed according to the requested frequency.

Place switch (2) in an upward position.

In order to be sure that the model transmitter operates correctly

place switch (3) and (4) downward and switch (5) upward.

Then turn knob (10) to the left until the pointer of voltmeter III is on the red line.

Check if the pointer of voltmeter IV is in the red zone.

Otherwise: the Modeltransmitter is not synchronized.

The Modeltransmitter transmits a high frequency wave to which low frequency is super-imposed.

In order to obtain good synchronization between high and low frequency place switch (4) in upward position.

Check if the pointer of voltmeter III is within the black zone.

Remember: At deviation adjust the torque with screw driver through opening (12).

Place Switch (4) downward again.

Place Switch (3) upward.

Turn Switch (7) to the left and right until you reach the maximum vane deflection and then return it to the zero position.

Remember: At deviation adjust the torque with a screw driver through opening (11).

If you know nothing about a transmitter
you may easily turn the wrong knob.

TESTING THE GROUND BEAM INSTALLED ABOARD THE AIRCRAFT

In the Launch Control Car

The central panel is installed in the Launch Control Car From this the Guide Beaa aboard the aircraft is tested.

The B -Frame

You must first operate the following Switches in order to test the B-Frame from the Launch Control Car

Turn Knob (10) to the left until you reach the maximum vane deflection.

Place Switch (4) and (9) in an upward positi«n.

Switch (5) in a downward position.

Switch (6) to the left.

AFTER COMPLETION OF THE TEST

On the B-Frame: Place Switch (l) and (2) in a downward position.

On the A-4: Remove the HF-Cable from the antenna red 2 or 4.

It is the miracle of electricity
MORAL: That makes it work, although we often don't understand it.

MOTTO: By testing you will know
Whether or not it is the way it should be.

By utilizing the compass you know whether you are heading in the right direction.
By utilizing the Guidebeam-Controlling apparatus, you know if the Guidebeam is heading toward the target.

1. You will direct a 25 meter high Dipolmast, then
2. you will ready the controlling apparatus, seperat.

Very important: Is the fact that only when launching with Guidebeam will you set up the device.

The entire controlling apparatus plus the Dipolmast is stored in the trailer.
Park the trailer in the designated spot.
Set it up so that the tongue of the trailer heads in the direction of fire.

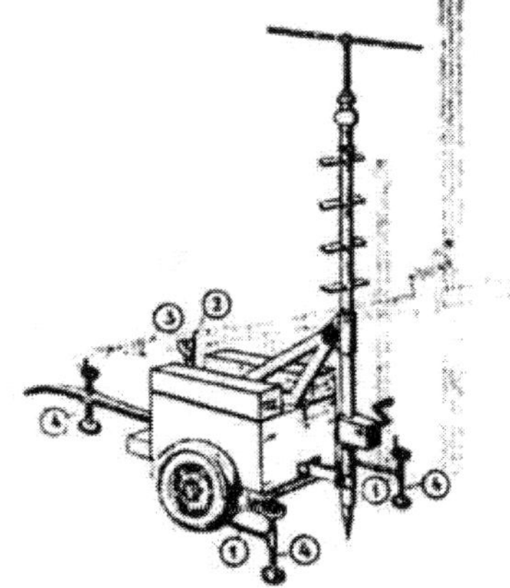

ERECTION OF THE DIPOLMAST

(1) Loosen the retainer bolts of the cantilever bolts. Twist the cantilevers toward the outside and backwards and stabilize them by bolts.
(2) Loosen the mast support.
(3) Remove the extension tube from the mast into the horizontal position.
Insert the extension tube into the bore of the uppermost mast tube, and fasten it with a clamp. By twisting them to the right. Place the Dipolmast Arms to the head of the extension tube.
Pull the inner extension tubes of the Dipolarm toward the outside until a marking becomes visible and screw them tight.
You have hereby made the correct arm adjustment.
Erect the mast horizontally and lock the back mast support.

Remember: When erecting the mast do not touch the extension tube.
The mast foot should be located directly over the marking.

(4) Twist the ground pegs in such a way that the mast will be horizontal.
The clinometer on the mast will guide you.

BRACING THE MAST

The span ropes are fastened to 8 ground stakes. You will arrive at the required points in the following way:

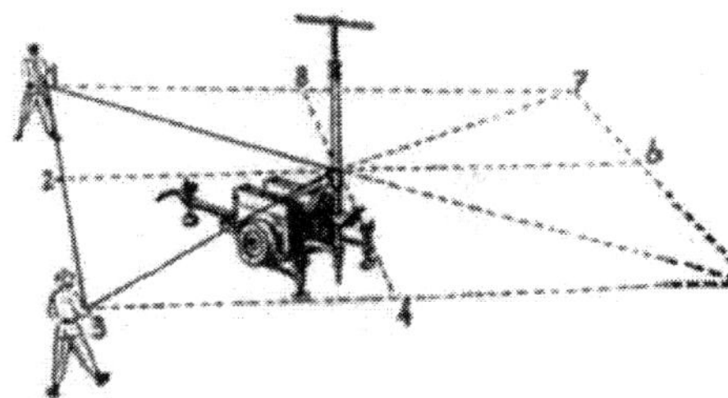

Place the large hook of the measuring tape over the mast pivot bearing around the mast and roll off the measuring tape in the same direction the trailer was travelling. As soon as the first splice with ring appears one man will hold the ring and remain at that particular point. You have hereby found point (1). Continue to walk to the left with the measuring tape until the 3rd ring appears. This ring will be held by the 3 man.

Roll off the rest of the measuring tape in direction of the mast. Hook the small hook at the end of the measuring tape into the ring of the large hook.

The 3rd man tightens the measuring tape. He is standing at point (3); the 2nd ring is at point (2). Drive in the stakes at point (1) and (3) for the highest and second highest mast bracing. Drive in the ground stake at point (2) for the two lower bracings.

Guide the measuring tape around the mast without removing the hook until all 8 points for the ground stakes have been established.

Drive in all ground stakes and unfold the steps for climbing the mast.

Connect the antenna cable to the plug of the extension tube.

Place the antenna roll on the unceiled axis.

Place the gray, blue, red, and green labled span ropes from bottom to top into their respective loops on the telescopic tube.

All span ropes are belaid out in a straight line leading toward the ground stakes and hook the guide drum on the stakes.

Tighten the 4 lowest guy ropes under the utilization of the clinometer.
Remember: The mast must always be horizontal. The rearward chain must always be unhooked. Wind out the mast. While doing this watch the guy-ropes continously.

Fasten the guy-ropes to the ground stakes without leaving any slack.
Remember: The strength of the mast depends upon just how accurate its horizontal position is.

In normal weather it is possible to wind out the mast without utilization of the guy ropes.

Should you wind out the mast in strong wind, then whatever side is exposed to the wind must be corrected by utilizing the respective guy ropes. The mast is horizontal if the bracing is correct.

PREPARING THE CONTROLLING APPARATUS

Check if the HF-Amplifier and the Ultra-high Frequency receiver is installed according to the ordered frequency. Connect the end of the HF-Antenna cable to the Amplifier inlet.
Check if the HF-Cable of the HF-Amplifier is connected to the Ultra-high frequency receiver.
Connect the FF-Cable to the respective coupling.
You have hereby made a connection between the Guidebeam-Ground device and the Guidebeam-Controlling apparatus.
Check if the battery is connected.

OPERATING THE CONTROLLING APPARATUS

By depressing the operating button (1) connect the Transformer blue. to the controlling apparatus. Turn the stage selector Switch (2) on blue. Regulate the battery voltage for the Transformer with the Regulator (3). The pointers of the Measuring device (3) must be in the black zone. After 5 minutes turn the Stage Selector Switch (2) on red. The pointer on the Measuring device (4) must be in the black zone.

<u>Otherwise</u>--The Ultra-high Frequency Amplifier will not have the correct one-voltage potential.

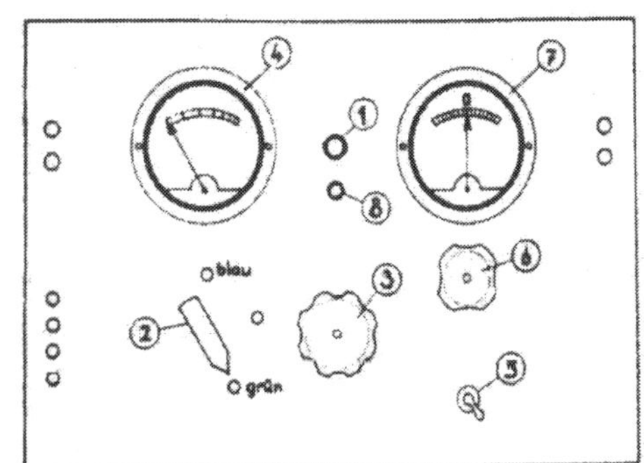

Turn the Stage Selector Switch (2) on green. The pointer on the Measuring device (4) must be in the black zone.

<u>Otherwise</u>--The HF-Amplifier will not have the correct anode-voltage potential.

Place the Transformer (5) on "loop-test".
Turn Knob (6) to the left and to the right.
Watch the Measuring device (7): Its pointer must travel beyond the entire scale area.
By utilizing Button (6) you check to see if the FF-Cable is connected with the Transmitter of the Guidebeam apparatus.
Check by phone if the Measuring device (12) shows a deviation (see Page'147).
Place transformer (5) on "measure".

<u>Remember:</u> Do not leave Transformer (3) on "loop test" for a longer period of time.

<u>Otherwise:</u> You run the battery down for no reason.

Depress Button (8).
You have hereby disconnected the controlling apparatus.

If all goes well at checking time
MORAL: Your missile will fly correctly.

The Rifleman loads his weapon and waits for the order "Fire".

It is not that easy with the A-4.

The Launching Platoon readies the A-4 for the Takeoff. Current produced in the current supply trailer, and led to the device by way of Cables, is needed for this purpose.

Functioning is controlled by electric Devices located on three desks in the Launch Control Car. To operate these desks is your business.

For launching you will only turn one switch.

Remember, however, when you do this you release a force of 25 tons.

Thanks to moto vehicles, changing
MOTTO: A position has finally become a joy

Two hundred years ago a General, who led his troops into battle, stood on top of a hill, which gave him a good view of the battleground. By watching the fighting he was able to determine the statue of the battle. Messengers on horseback delivered his orders.

With the A-4 it is not possible to determine the working order of the numerous devices by simply watching them. At the launching installation you therefore need special vehicles equipped with special instruments in order to prepare and test the A-4 for launching. Orders are transmitted by cable.

The following makes up a launching installation:

1 Launching control car

1 Test car

1 Current-supply trailer

You are to drive these vehicles on to the launching site and operate them.

Park them as ordered.

Remember: You must have a clear view from the launch control car to the launching platform.

In the Launch control car is located the technical Command Control Station at the launching site.

In the Test Car certain instruments are installed enabling you to test the radio devices of the A-4 prior to takeoff. How this is done is shown on Page 69 to 72. Protect the device from dust and dampness.

In the Current-Supply trailer 220/380 V Polyphase current and 27 V Direct current are produced. In addition you are also supplied with Battery voltage.

Move up fast and camoflage well
MORAL: Otherwise you will soon be bombed

MOTTO: It is a well known fact that laying a cable
Is connected with movement.

As soon as the launch control and test vehicles are set up in the launching site, you are to connect the cables.
Remember your buddies at the A-4 are unable to continue until current flows through the cable.

PRIOR TO LAYING THE CABLE

Assemble the cable car: The chassis is stored at the rear wall, the wheels at the outside wall of the current supply trailer.
Take the cable drums from the cable trailer of the current supply trailer. Additional cable drums may be found inside the current supply trailer.

LAYING THE CABLE

Take into consideration that the cables must be protected.
Wherever vehicles cross the cable you must
either bury the cable, or
cover the cable with planks.

Protect the fittings from dirt and dampness.
Each cable carries a number and a marker at the end. The sockets carry the same numbers.
The marker tells you where to connect the cable. Pay attention to the numbers and the markers when making the connections.

Otherwise--you will cause a short circuit.

Connect the following equipment between each current supply trailer and its respective long range Rocket-Carrier:
Two 60 m anti-aircraft cable 108 X 0.5^2
One 60 m NSH-Cable 21 X $1,5^2$
One 60 m NSH-Cable 4 X 16^2.

Count the following equipment between each current supply trailer and its respective launching platform:

One 70 m NSH-Cable 21 X 1.5^2.

Connect the following equipment between each current supply trailer and its respective launch control car

One 120 m anti-aircraft cable 108 X 0.5^2 and
one 120 m NMH Cable 4 X 4^2.

From each test car the following is connected to the respective power set

one 70 m NSH Cable 21 X 1.52 and
one 70 m NSH Cable 4 X 2.52
to each launching platform:
one HF-Cable.

COMMUNICATION CABLES

You will also see to it that there is a means of communication between the individual launching sites. You must therefore hook up telephone wires. You will find them in the cable drum and current supply trailer.

From the launching site connect
two 180 m FF-Cables to the respective launch control car.
From the current supply trailer connect
one 70 m SF-Cable to the respective long range rocket-carrier.
From the test wagon connect
one 40 m SF-Cable to the respective long range rocket-carrier.
You will find the 40 m SF-Cable in the test cars.
70 m SF-Cable in the individual current supply trailer.

As soon as the test cars are properly parked and set for operation, you must reconnect them by utilizing a 150 m SF-Cable, but this time you connect it with the launch control car.

Should the Guidebeam be used for launching, then reconnect.
one 500 m FF-Cable from the launch control car to the Guidebeam control device.

Should an emergency arise due to a damaged FF-Cable you may substitute a normal 250 m cable obtainable from any Signal outfit.

Remember: Under no circumstance are you to utilize Pupincoil.
Otherwise: --Your communication system will fail.

MOTTO: When laying your cable protect your fittings from dirt and dampness

MOTTO: Everybody agrees that it is a fine thing to be able to produce your own electricity.

THE CURRENT SUPPLY TRAILER

The current supply trailer harbors an auto-engine that actuates a polyphase current generator of 220/380 Volts.
After receiving the order, you will immediately start up the generator set.

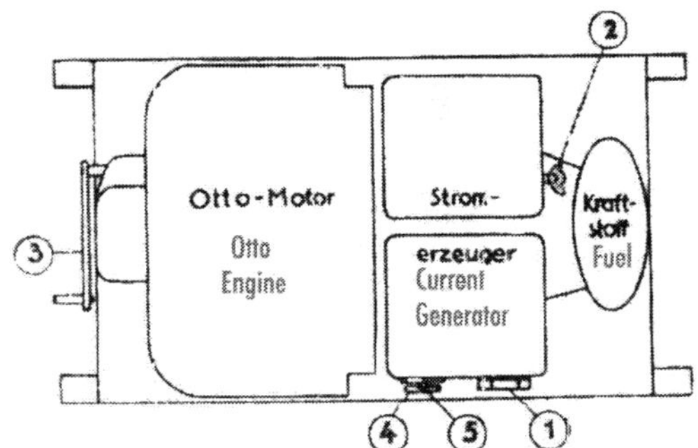

(1) See that the line is connected.

(2) Ground the generator. The red-wing nut at the foot of the generator underneath the fuel tank is the earth clamp.

(3) Turn on the auto-motor. As soon as it runs place the ignition switch of the auto-motor on "operate".

(4) Set the voltage by turning the hand regulator of the switch-box without placing a stress upon the generator. At a stress you are not allowed to make a re-adjustment.

(5) Turn on the main switch at the switch box. Your generator will then produce current.

DISCONNECTING

Turn the carburator throttle lever on iding.
Turn off the main switch located on the switch box.
Depress the stop button on the front side of the auto.motor until the motor stops.
Do not forget to turn off the fuel tap.

CHECKING THE CONTROLS
After receiving the order turn the pole of the cut-off signal modulator battery "on", or as the case may be "off".

MORAL: Should the meter not be running
The lamp will not be lit.

MOTTO: This vehicle is protected with armor and
is the switch-control station for the launching site.

Should you be stationed in the launch control car, then you are in the technical command control station of the launching site.

Never forget this when working there.
The most important things in the launch control car are the three control desks, viz:

The propulsion control desk

The piloting desk

The radio control desk

The X-time plan gives you the switching sequence of the individual desks and how they work into each other.

Your job is as follows:

THE PROPULSION CONTROL DESK

It is utilized for watching over propulsion and launching.

After receiving the order ready yourself for propulsion control testing.

From the Launching Platoon Leader

You will insert the firing key and turn it on "testing".

Check if the condition signal lights for "testing" "ground power supply", "Plug I" and "Plug II" are lit.

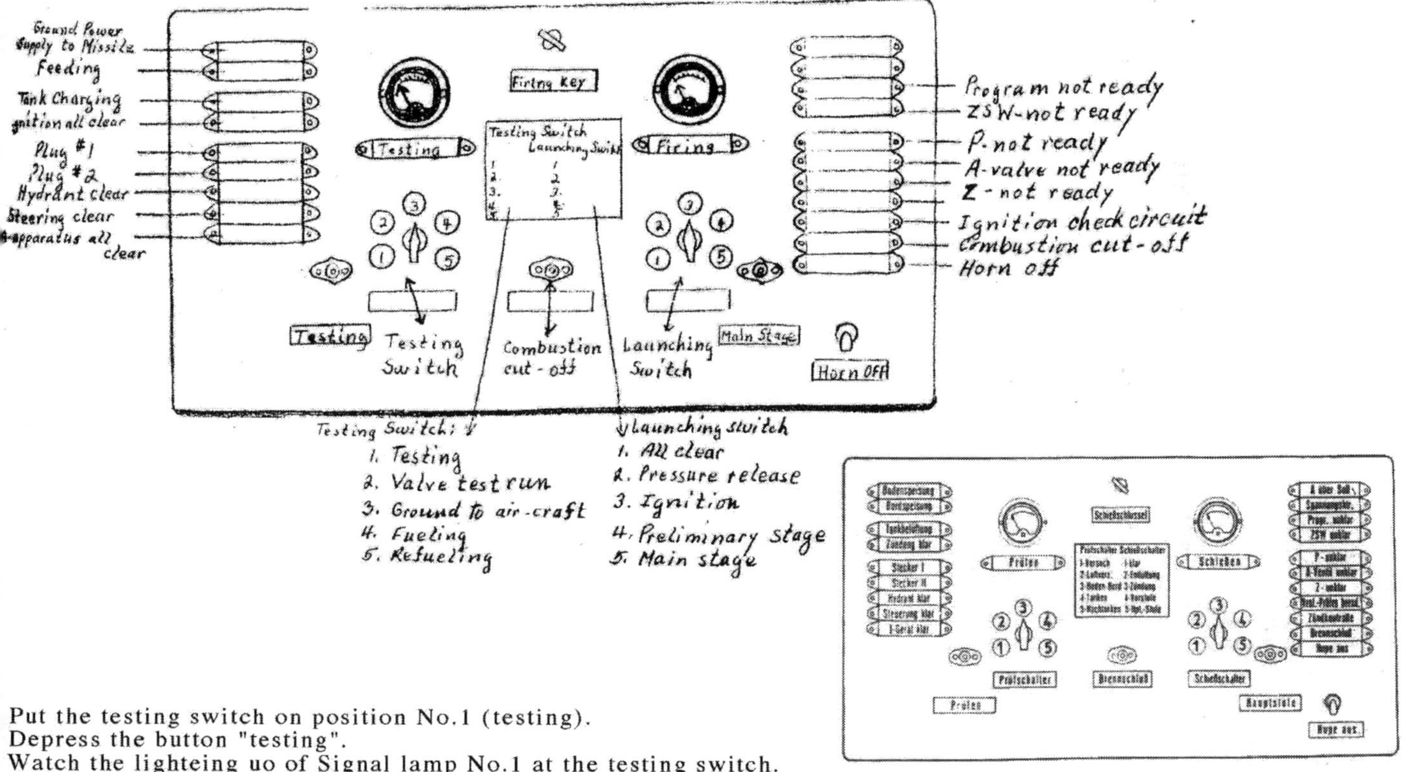

Put the testing switch on position No.1 (testing).
Depress the button "testing".
Watch the lighteing uo of Signal lamp No.1 at the testing switch.
Should a "not ready" lamp light up, then report it to your platoon leader.
Now report "Ready for propulsion unit testing."

To the Order:

1. <u>Switch the launching valve to the Position 1--"all clear"</u>
   Watch the lighting of the "all clear" signal light 1 on the launching valves.

2. <u>Switch the launching valve to Position 2--"pressure Release"</u>
   Check if the condition signal light "Tank pressurizing" is lit and if it goes alternately on and off.
   Watch the lighting up of the "all clear" signal light 2 on the launching valve.

3. <u>Switch the launching valve on position 3--Ignition</u>
   Check if the condition signal light "Ignition clear" goes off.
   Check the lighting of the signal light 3 on the launching valve.

4. <u>Switch the launching valve on Position 4--Primer</u>
   Check the lighting up of the signal light 4 on the launching valve.

5. <u>Turn off all switches</u> Remove the firing key. Turn the firing key back on Position 1.

After the launching platoon leader gives the order:

<u>Ready for "ground to aircraft testing</u>".
Insert the firing key and let it stay on "testing".
Check if the condition signal lights light up on "testing", "gorund power supply", "Plug I" and "Plug II".
Place the testing switch on Position 3 (ground to aircraft testing).
Report: "Prepared for ground to aircraft testing".
Depress the test switch after receiving the order to do so.
Check if the "ground power supply" light goes off and the "aircraft power supply" goes on.
After receiving the order "Turn off all switches" you will place the test switch back in its old position
And remove the test key.

<u>Prepare for the main test run</u>
Insert the ignition key and turn it on "launching".
Check if the condition signal lights and the "all clear" lamp "launching" ground power supply "Plug I", "Plug II" "Hydrant clear", "Steering clear" and J-Device light up.
Report: "Ready for main test run".

After receiving the order:

1. Turn the launching valve on Position 1--Clear
   Watch the lighting of the "all clear" signal light 1 on the launching valve.

2. Turn the launching valve on Position 2--Pressure Release
   Check if the condition signal light "Tank Pressurization" lights up and then goes off again.
   Check if the condition signal light "all clear" lights up.
   Check the lighting up of signal light 2 on the launching valve.

3. Switch the launching valve on Position 3--Ignition
   Check if the condition light "Ignition clear" goes off.
   Check if the light "Ground power supply" goes off and then the light "Aircraft power supply" goes on.
   Check the lighting up of signal light 3 on the launching valve.

4. Switch the launching valve on position 4--Primer
   Check the lighting up of signal light 4 on the launching valve.

5. Switch the launching valve on Position 5--Main Stage Depress the lever "Main Stage".
   Check if the condition signal lights "Plug I" and "Plug II" go off.

6. Turn off all switches--Remove the ignition key--Turn the launching valve back on Position I. After receiving the order "Prepare for Fueling"

Insert the firing key and turn it into the Position "Testing".
Check the condition signal lights "Testing", "Ground Power Supply" "Plug I" and "Plug II" will light up.
Place the testing switch on Position 5.
After receiving the order, you will depress the lever marked "Testing".
Inform the launching platoon leader as soon as lamp A lights up.
After receiving the order turn off all switches.
Place the test switch into its original position and pull the ignition key.

LAUNCHING

After receiving the order you will follow the same procedings you did in the "Main Test Run".
Check in addition if the control lamp A lights up at Position 1.
Wait at Position 2 until the lamp goes off.

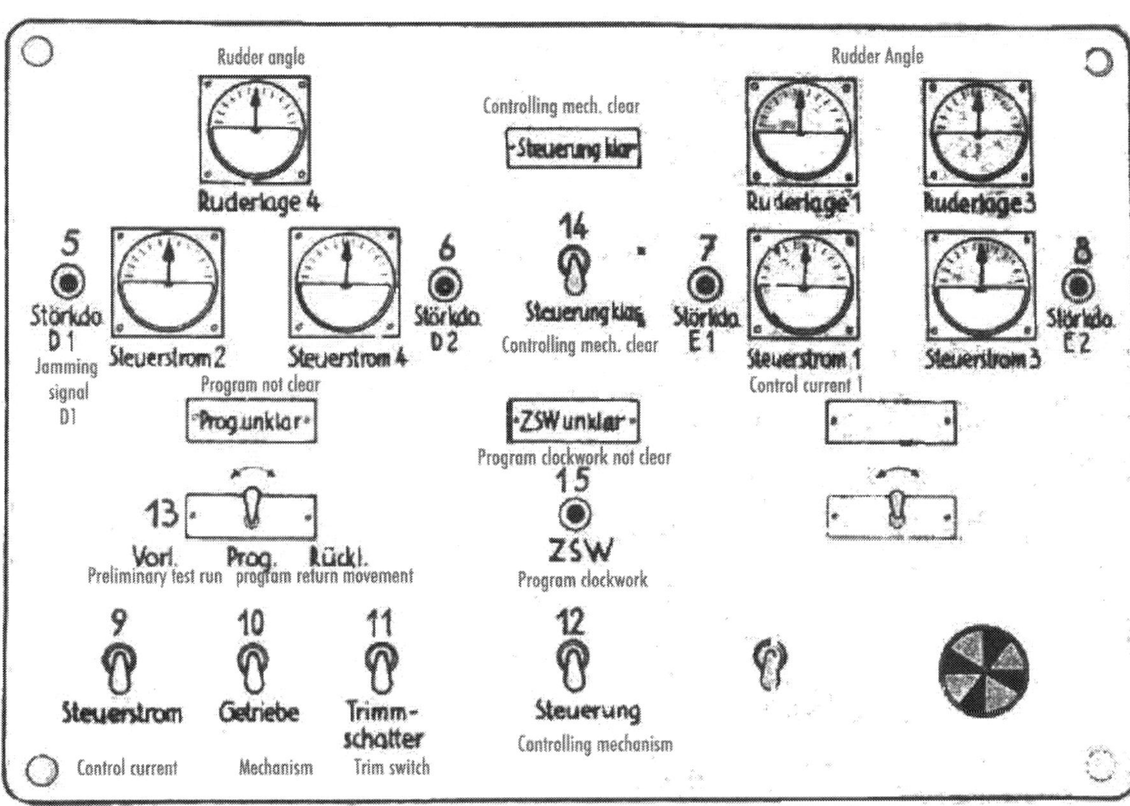
Rudder angle
Controlling mech. clear
Steuerung klar
Rudder Angle
Ruderlage 4
Ruderlage 1
Ruderlage 3
5
Störkdo.
D 1
Jamming
signal
D1
Steuerstrom 2
Steuerstrom 4
6
Störkdo.
D 2
14
Steuerung klar
Controlling mech. clear
7
Störkdo.
E 1
Steuerstrom 1
Control current 1
Steuerstrom 3
8
Störkdo.
E 2
Program not clear
Prog.unklar
ZSW unklar
Program clockwork not clear
15
ZSW
Program clockwork
13
Vorl.
Prog.
Rückl.
Preliminary test run
program
return movement
9
Steuerstrom
Control current
10
Getriebe
Mechanism
11
Trimm-
schalter
Trim switch
12
Steuerung
Controlling mechanism

THE CONTROL DESK

It is used for the following: 1. In order to clear the controls, 2. In order to test the Guide-Beam installation aboard the aircraft.

CLEARING THE CONTROLS

In order to check the steering the ground voltage must be connected and the A-4 must be in a horizontal position.

Adjusting the controls

Turn on Switch (9) and (12). Switch (12) starts the Transformer 1 and 2.
Check if Ammeter 1,2, 3, and 4 show different values and turn on Switch (10). Ammeter 2 and 4 must operate on the same movement. Then turn off Switch (10).

Give the following order to the current trailer: "Disconnect Course Gyro-battery!"
The pointer position 1 to 4 will now change somewhat.

Give the following order to the Launching Site: "Rectify control current 1!"
Give the order "Stop!" as soon as the control current matches the value zero.
Do the same for control currents 2, 3 and 4.

Give the order: "Connect the Course Gyro-Battery!" to the current supply trailer.
Should the control current 2 and 4 not be zero, then give the order: "Adjust Horizon!" to the Launching Site or (working platform).
Give the order "Stop!" as soon as the control current 2 and 4 is on zero.
Should control current 1 and 3 not have the same pointer position or not be on zero then give the order "Adjust roll and yaw gyro!"

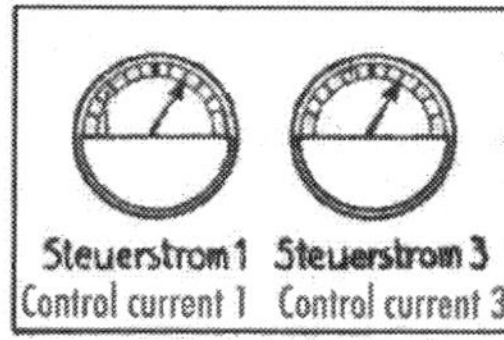

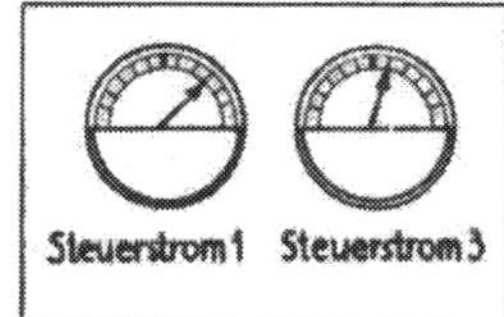

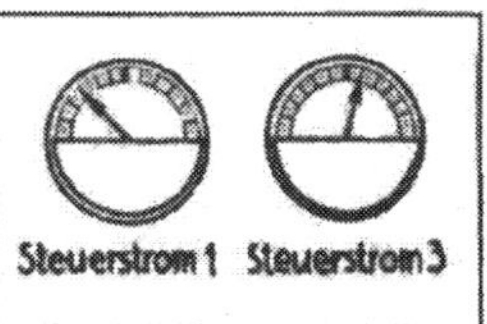

Give the order "Stop!" as son as control current 1 and 3 show the same pointer Position.

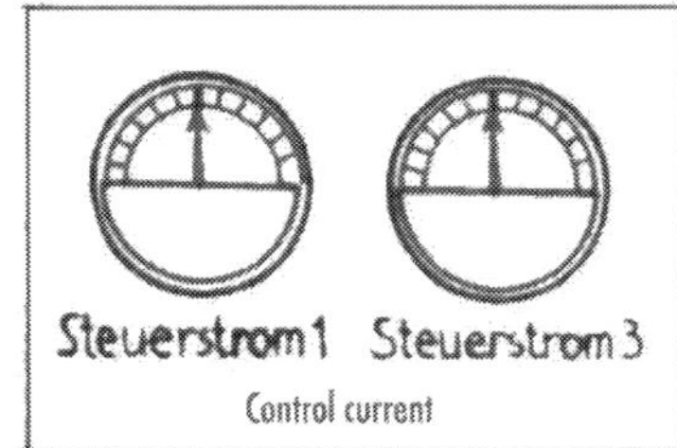

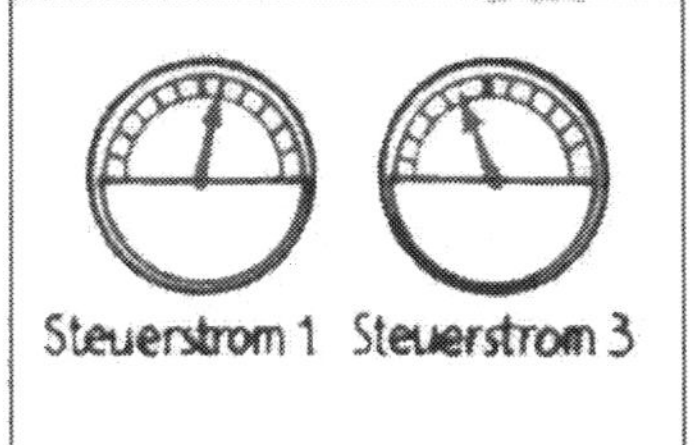

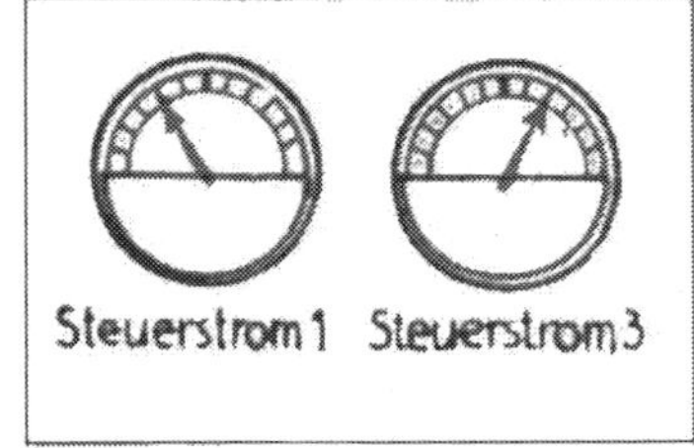

Again you must rectify control current 1 and 3, should they be larger than O, but smaller than 4 and A. Should the control current 1 and 3 be over 4 and A, then you must interchange the roll and yaw gyro.

SERVO MOTORS AND SYNCHRONIZATION

Turn on Switch (10).
Give the order "Check synchronization of Rudder 2 and 4".

Depress interference Button (5) (D1).
Check if Ammeter 2 and 4 show the same pointer movement to the Left.

Depress interference Button (6) (D 2).
Check if Ammeter 2 and 4 show the same pointer movement to the Right.

Depress interference Button (7) (E 1).
Check if Ammeter 1 and 3 show the same pointer movement to the Left, and if missile angle Meter 1 and 3 go to the Left end position equally fast.

Depress interference Button (8) (E 2).
Check if Ammeter 1 and 3 show the same pointer movement to the Right, and if missile angle Meter 1 and 3 go to the Right end position equally fast.

In order to check synchronization of collaborating servo motors give interference signals to the Right and to the Left with short time intervals.

Check if the missile angle Meters 1 and 3 show the same speed of movement.
Should this not be the case, give the order "Rectify Potentiometer of Rudder 1 and 3!"

TESTING THE PROGRAM

Make Rudder 2 and 4 deflect toward Fin 3.
Check if missile angle Meter 4 makes a movement to the Left.

Turn Transformer (13) to the Left and count aloud 21, 22, 23, 24.
The indicator "Program not clear" must light up.

After 4 seconds the Ammeter and Missile Angle Meter 2 and 4 must go to the Right.
Turn off Switch (10).

Place Transformer (13) to the Right and 3 seconds later place it on the middle position.
After 2 minutes the indicator "Program not clear" must go off.

TESTING THE TRIM STEERING

Turn on Switch (10) and (11).

Give the order "Correct the deviation of Rudder 1 and 3!" and then check the path of the Trim Tabs 2 and 4.
Missile angle Meter 1 and 3 must show a contra-deflection.

After you have reported "Trim Tabs run correctly!"
Turn off Switch (11).
After reporting "Trim Tabs reverse correctly!" check if the two indicators "Program not clear" and "Contact Timer not clear" have died out then make the report "Controls clear" to the officer in charge.
Turn on Switch (14).

THE STEERING MECHANISM

The steering mechanism is disconnected only if, for special reason, a longer interval prior to the main Test run should be necessary.
Turn off Switches (1), (10) and (12).

KEEPING THE SERVO MOTORS WARM

This is necessary only in very cold weather under 5° outside temperature.
At regular intervals turn on Switch (10), after 3 to 5 minute turn it off again.

Otherwise-the oil will become thick.

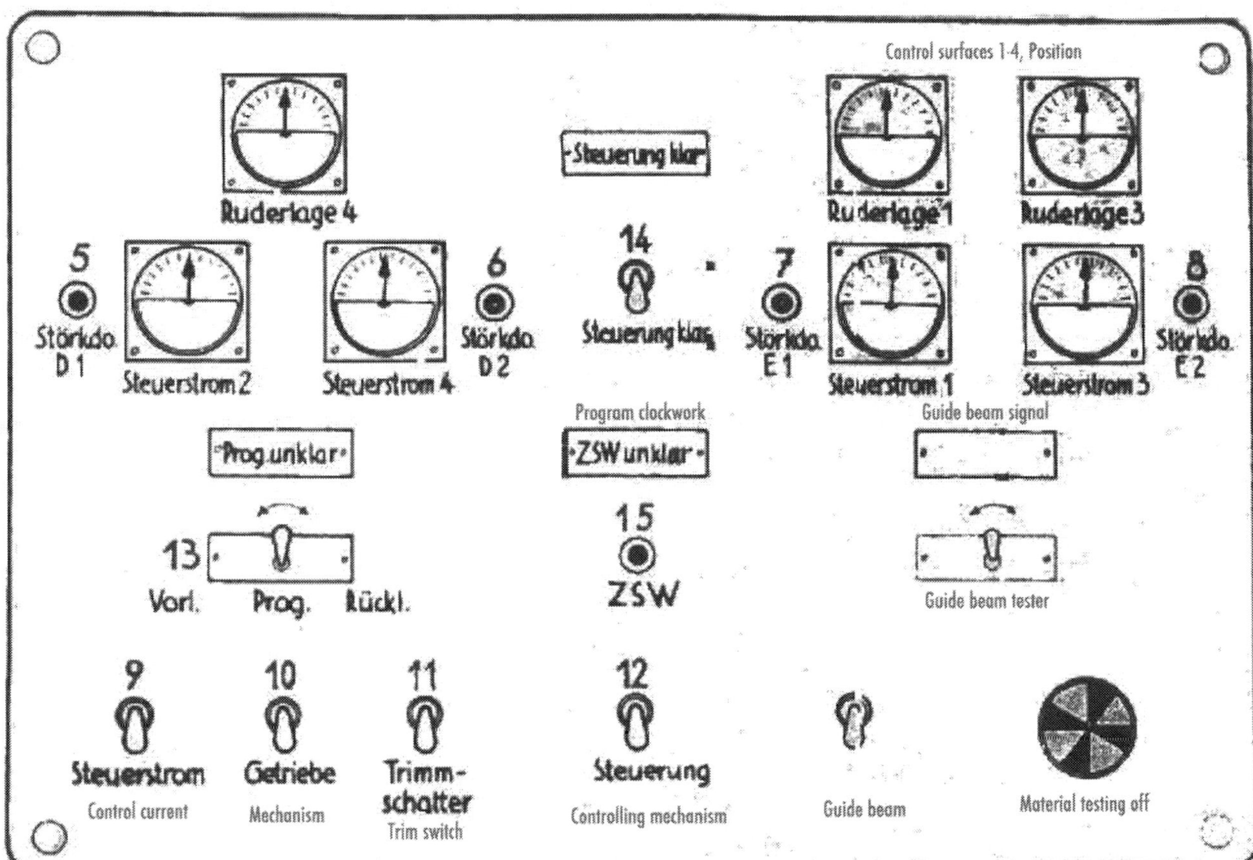
Control surfaces 1-4, Position
Steuerung klar
Ruderlage 4
Ruderlage 1
Ruderlage 3
5
Störkdo.
D 1
Steuerstrom 2
Steuerstrom 4
6
Störkdo.
D 2
14
Steuerung klar
7
Störkdo.
E 1
Steuerstrom 1
Steuerstrom 3
8
Störkdo.
E 2
Program clockwork
Guide beam signal
Prog.unklar
ZSW unklar
15
13
Vorl.
Prog.
Rückl.
ZSW
Guide beam tester
9
10
11
12
Steuerstrom
Control current
Getriebe
Mechanism
Trimm-
schalter
Trim switch
Steuerung
Controlling mechanism
Guide beam
Material testing off

THE MAIN TEST RUN

Should the steering mechanism become disconnect:
Turn on Switch (9) and (12).
The gyros start up, the Control current oscillates and then goes directly on zero.
If necessary turn on Switch (10) for a short time.
As soon as all Control currents are zero and you are sure that the indicators "Program not clear" and "Contact Timer not clear" do not light up, you will report to the Battery officer: "Steering Mechanism clear for main test run".
The still detectable small false currents must keep their course.
Upon receiving the order "Take off":
Watch Ammeter 2 and 4. After 4 seconds they are to go off to the Right.

Disconnecting the control switches

Turn off Switches (9), (10) and (12).
After inserting the Magnetic plugs turn Transformer (13) to the right for a short period of time and depress Button (15) ZSW for a few seconds.
After approximately 1 minute the Indicator "Program not clear" will die out. It is followed by the indicator "ZSW not clear". (After 2-1/2 minutes).

Keeping the Servo Motors Warm

Turn on Switch (10) at regular time intervals. After 3 to 5 minutes turn the switch off.
In cold weatherthis is done immediately after fueling with A-substance, in warm weather fill up with A-substance first.

CLEARING PRIOR TO LAUNCHING

Takes place after fueling and erection, as well as turning the A-4 in the direction of fire.
Turn on Switch (9) and (12).
Wait 3 minutes
Watch the Control currents.
Should it be necessary, rectify them again.

TESTING THE GUIDE BEAM-APPARATUS ABOARD THE AIRCRAFT

Putting Into Operation

Place Switch (9), (12) and (16) to the top.
The indicator (17) "ZSW not clear" should not light.

Otherwise—the contact timer will not be in the zero position.

Indicator (18) "Data Transformer" must light up.

No signal will get to the control amplifier.

Remember: If Indicator (17) "ZSW not clear" lights up, depress Button 15 approximately 18 Seconds. Indicator (17) must then die out after 1 1/2 Minutes and Indicator (18) must light up.

TESTING THE CONTACT TIMER

Depress Button (15) for approximately 8 Seconds.
Indicator (17) must light up.
Indicator (18) will die out.

Otherwise-no signals will be transmitted from the Guide Beam Device

Testing the Guide Beam Signals

Place Switch (19) "Guide Beam Tester" on the middle position.
Control currents 1 and 3 may amount to + 10 mA.

Remember: Oscillation of control current is permissible.
Turn Switch (19) to the right.
The control current measuring device 1 and 3 must first deflect strongly to the right, then return to just a partial deflection.
Place Switch (10) "Transmission" to the top.

Otherwise-the Rudders will not be able to move out in accordance with the control current.

See if the "Position Indicator" 1 and 3 deflect to the right.

Place Switch (19) to the left.

The Control Current Measuring Device 1 and 3 must deflect strongly to the left and then return toonly a partial deflection.

See if Position Indicators 1 and 3 deflect to the left.

Place Switch (10) to the botton.

Depress Button (15) ZSW approximately 10 Seconds.
The Control Current Measuring Device 1 and 3 must slowly continue to deflect to the left.

Otherwise--the A-4 will not steer correctly at a cross wind.

After 1 1/2 Minutes the Transperancy must die out and Indicator (18) must light up.
Check if the Control Current Measuring Device 1 and 3 return to zero.

Otherwise--the Signal Mixing Motor and the Measuring Data-Transformer did not return to zero.

Place Switch (19) in Middle Position.
Depress Button (15) approximately 18 Seconds.
Transperency (17) must light up and Indicator (18) die out.
The Control Current Measuring Device 1 and 3 may show an average of 25 mA at the most.

Otherwise--you will be forced to interchange the Guide Beam Device aboard the aircraft.

After 1 1/2 Minutes Transperency (17) must die our and Indicator (18) light up.

The Control Measuring Device 1 and 3 must return to zero.

AFTER COMPLETTING THE TESTS

Place all Switches on the Control Desk in a downward position.

MORAL: Once you are finshed with testing the Guide Beam, the next step will be to study combustion cut-off.

RADIO CONTROL DESK

The Radio Control Desk is utilized for testing the Combustion Cut-off Device, that is

either the Frequency Dopplers and Cut-off Signal-Receivers, if combustion cutoff is given via radio,

or the J-Device. Theintegrator Testing Device Calibrator JS 1, located in the center part of the radio control desk is utilized for testing the J-Device 1/1--3

Page 74 will tell you how this center part looks and how you are to operate it.

The right and left part of the Radio Control Desk is utilized for connecting the Frequency Doppler and testing the Cut-off Signal Receiver.

ENGAGING THE FREQUENCY DOPPLER

The test may begin as soon as the Magnetic Plugs are connected.

When given the go ahead connect transformer No.3 in the A-4 by engaging tumbler switch (1) "Hydrant".

TESTING THE CUT-OFF SIGNAL-RECEIVERS

The magnetic plugs must also be connected for this test.

Contact the test car by phone and give the following order:

1. Turn the Switch in Field III (Transmitter red) on "in".
2. Turn the Switch in Field II on "in".
3. Turn the switch in Field 22 (Cut-off Signal Modulator) on "in".

Wait 1 Minute until the Apparatus has warmed up.

a) Unarmed Test

Turn on Tumbler Switch (1) "Hydrant". Transformer 3 will start up. Wait 1 to 2 minutes. Depress the "Cutoff Signal Modulator" push button.

After a short time the following must appear:

1. Shutter Enunciator Signal 1a-(Preliminary cut-off signal)
2. After approx. 3 seconds Shutter Enunciator Signal 1b (Main cut-off signal).

b) Armed Test

Depress both push buttons "cut-off signal modulator" and "Armed Signal" simitaneously. The following must appear:

1. Shutter Eununciator Signal 1a. It stops.
2. After approx. 3 seconds shutter enunciator "signal 1b" and shutter enunciator "signal armed". At the Propulsion Unit Control Desk the indicator "Combustion Cut-off" must light up. At this time the siren must start up.

Remember: Should you get interference from the Cut-off Signal Receiver, then do not permit the transformer to be connected for more than 15 minutes.

Should everything be all right report "Combustion Cut-off clear".

Turn all Switches on "off".

MAIN TEST RUN

Five minutes prior to Test Begin, turn on Tumbler Switch (1) again.

After receiving the order "Combustion Cut-off" depress the two push buttons "Cut-off Signal Mosulator" and "Live Signal" located on the radio control desk. You have hereby by way of the HF Cable 4 transmitted the modulated energy of the test transmitter to the cut-off Signal Receiver.
The Indicator Receiver 1a, Signal 1b and Live Signal must appear in the same sequence as shown in the paragraph "Live Test".
After a flawless test reoprt:
"Combustion cut-off Device aboard aircraft ready for Launching".
After completing the test turn Tumble Switch (1) on "off".

LAUNCHING

Turn on Tumble Switch "Hydrant".

MORAL: "Main Stage--everybody is waiting to see if the A-4 starts up correctly.

MOTTO: Even though a job seams small and minor
It is very important for effectiveness.

NOSEPIECE OF THE FUSE

The Elephant utilzes his trunk as a Sense of Touch.

By utilization of its Nosepiece the A-4 will know,
when it touches the Ground.
It burts into thousands of pieces;
the Explosive Charge explodes.

The Nosepiece of the A-4 harbors the fuse; upon impact it must ignite the explosive charge in less than a thousandth of a second.

<u>Otherwise</u>--the A-4 would be driven deep into the ground.

How is the Fuse armed?
Screwing in the Fuse does not arm it.
During testing of the Power Plant it remains unarmed.
Nothing therefore can happen.
The Fuse is armed in three Steps and this is only after launching of the A-4.

<u>Arming the Fuse takes place automatically during flight.</u>

After receiving the order place the Communication Apparatus plug into the Sterg (code for a type of fuse-arming unit in Guided Missiles) which delivers the ignition voltage. Connection to the Sterg can be made through III of the Instrument Compartment.

After receiving the order remove the covers from the two fuse Connection Plugs and place the plugs into the Fuse Sockets.

The Communication Apparatus plugs as well as the covers for the Fuse Connection Plugs are sealed. Remove the seals only prior to making your connection.

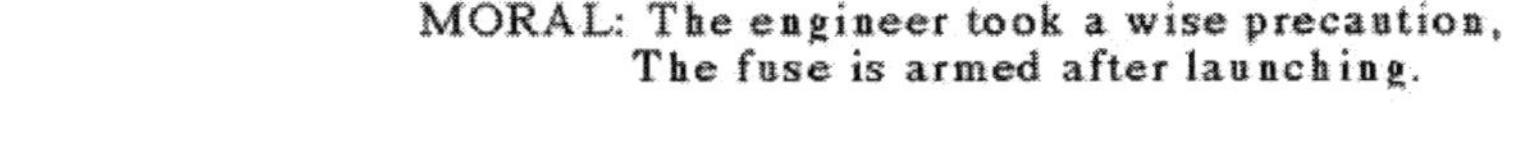
MORAL: The engineer took a wise precaution,
The fuse is armed after launching.

The A-4 derives the Power needed for Flying from the Fuel.

Over 8000 liters of Fuel must be filled into the A-4 prior to Launching.

You will haul the Fuel in Tank Trucks from the supply points to the launching site.

There you will fuel the A-4.

Handle the Fuel with care and abide by the Safety-Regulations.

MOTTO: Whether or not it is possible to accomplish something great depends on the fuel.

In order for the A-4 to perform such a long flight, you will fill it with four Liquid Fuels.
It will take a terrific Amount until it is filled up:
Each A-4 must have over 8000 liters.
Some fuels have a complicated chemical Name.
We will therefore name them:

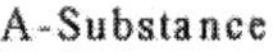

A-Substance
B-Substance
T-Substance
Z-Substance

You can remember them easily.
The gas of combustion is produced from the A-Substance and B-Substance.
Steam is produced in the generating plant by the T-Substance and Z-Substance.
What special precautions do you have to take when working with various Fuels?

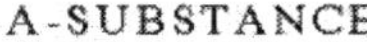

A-SUBSTANCE

Color pale blue.

A-Substance is very cols (-183° C). It changes dampness to ice.

<u>Therefore:</u> All pipe lines must be dry.

Otherwise—they will freeze.

Thawing with warm water or hot air brings about a loss of time.
The coldness of the A-Substance acts like Fire. Should A-Substance get on your skin it will cause blistering.

<u>Therefore:</u> Wear at least a pair of Gloves made of Asbestos.
If possible wear complete protective clothing.

A-Substance-Steam is also very cold.

<u>Therefore:</u> Wear warm Shoes, should it be necessary for you to stand in A-Substance for a long period of time.

Should you spill A-Substance on rubber tires they will break. Therefore: Wash off with plenty of water.

Burnable material saturated with A-Substance will ignite immediately by one single spark.
<u>Therefore:</u> Do not smoke! Do not use an open fire or open light!

Oil and fatty substance ignite spontaneously when mixed with A-Substance.
<u>Therefore:</u> Do not use oil or fatty substances!

A-Substance evaporates continuously.
<u>Therefore:</u> A-Substance containers must be open.
Otherwise—they will explode.
Speed up your A-Substance transports!
Otherwise--you will arive with empty containers.

B-Substance

Color blue or violet.
Just like ink.
B-Substance is very poisenous. One shot glass will blind you, Shot glasses will kill you.
<u>Therefore:</u> Do not drink one drop of B-Substanc

B-Substance is very inflammable!
<u>Therefore:</u> Do not smoke! No open fire or open

B-Substance should not be mixed with A-Substance and T-Sul
<u>Otherwise</u>—there will be danger of fi
explosion

T-Substance

Color from colorless to a slight yellow.

T-Substance eats up your skin and ignites your clothing. Be on lookout for any contact with the T-Substance.
<u>Therefore:</u> Wear protective clothing made of Mi
(a type of synthetic rubber)
Do not forget your safety goggles.
<u>Otherwise</u>--you will contact blisters or burns.

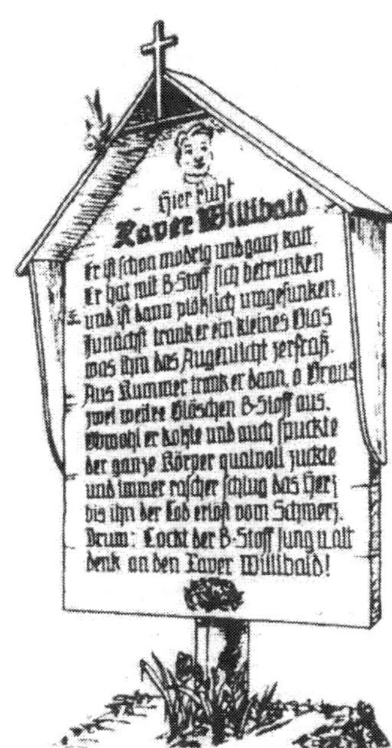

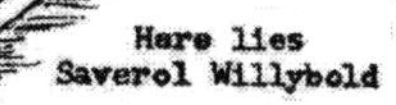

He is already mouldy and very cold. After intoxicating himself with B-Substance he suddenly collapsed. First he drank a small shot of B-Substance which caused him to lose his eyesight. Out of grief he drank two additional shots. His whole body shook painfully and his heart started to beat faster and faster until death finally relieved him of his agony. Should B-Substance tempt young and old then think of Saverol Willybold!

Keep water handy in order to wash off areas contaminated with T-Substance. The water tank located on the T-Substance Oil Tank Truck holds 300 Liters.

T-Substance affects metals also.

Therefore: Utilize containers made of glass or aluminum only.

T-Substance decomposes continouosly and gives off a gas. This happens especially when contaminated with outside matter.

Therefore: Cover all containers well. Keep hoses and mountings clean. The gas however must be able to emerge.

Otherwise:--the container will burst!

When T-Substance decomposes heat is produced.

Therefore: Check the temperature of the containers regularly by touching them.

T-Substance crystallizes at -19° C.

Therefore: Heat in cold weather with T-Substance-Preheater!

T-Substance must have a concentration of at least 78.4%.

Otherwise--the steam temperature will drop and the thrust of the A-4 will not be large enough.

Therefore: Check the specific weight by submerging an aerial meter. Read off the temperature and derive the concentration from a chart.

Put out T-Substance fires with water only, and never use the Tetra-Fire-Extinguisher on the truck.

Otherwise--you will increase the danger. A minor thing has often caused an explosion.

Z-SUBSTANCE

Z-Substance looks similar to blueberry soup. It is a deep dark violet color.

Z-Substance eats up your skin.

Therefore: Wear the same protective clothing as you do when working with T-Substance.

Z-Substance crystallizes easily.

Therefore: Always preheat the container of Z-Substance with the electric heating device. Prior to preheating open the lock of the container.

Otherwise:--it might hit you in the face.

Z-Substance should not be mixed with T-Substance.

Therefore: Always store Z-Substance and T-Substance separately. Never transport the two substances on the same vehicle. Never use the same container for storing.

Otherwise-- there will be an explosion.

MORAL: Fuel often causes grief
Even if you believe it to be stored properly.

It is to your advantage to keep your
MOTTO: "Eyes open" especially during fueling.

1. You will haul the fuel from the depot. You will fill the A-Substance, B-Substance and T-Substance in the tank truck. Z_Substance is transported in cans from the field supply depot.
2. You will drive the tank truck to the launching site.
3. You will fill the A-4 with fuel at the launching site. The trailer and propulsion unit crew will assist you.

B-SUBSTANCE

The container KW is utilized for transporting B-Substance from the supply depot to the launching site.

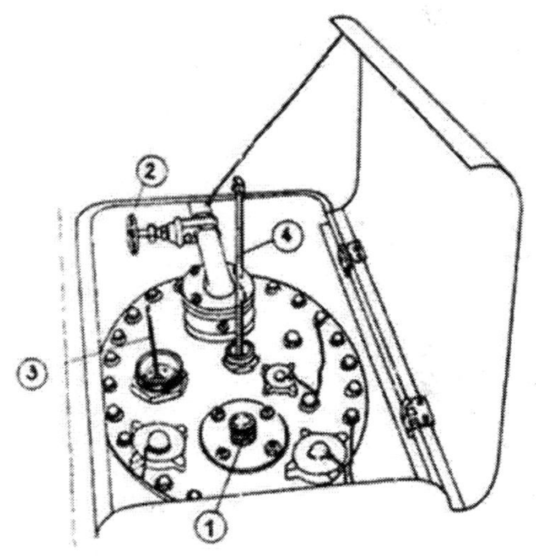

FILLING THE CONTAINER KW

You may fill up the B-Substance container
1. through the filler cap at the manhole cover
2. through the drain plug at the instrument compartment.

The instrument compartment is located at the rear of the truck.

<u>Remember:</u> Prior to the filling:
Unwind the screw-cap of the filler plug (1).
Connect the filler hose.
After filling:
disconnect the filler hose and screw on the screw-cap.

<u>Remember:</u> The pressurizing and pressure release valve (2) located inside the manhole cover must be open at all times.

<u>1. Fueling through the filler plug</u>

Turn the fourstage faucet in the instrument compartment on A or C; now you may start with fueling.

POSITION OF THE VALVES AT FUELING AND DRAINING OF B-SUBSTANCE

↺ Open the valve  Close the valve

The number will give you the sequence

	Process	Tank-Truck			Motor-Pumping Device			
		Drainage Plug	Four-way faucet	Cut-out gov. valve	Three-way faucet	Cut-out governor valve		
						1,2	3	4
Filling the truck	1. Case. Through Filling Sleeve	↻ (1)	A or C (2)					
	After filling	↻						
	2. Case. Through drain plug	↺ (5)		↻ (7)		↺ (1)	↻ (3)	↺ (2)
	A. Motor pump	↻ (4)	B (6)	↻	B (4)	↻ (1)	↻	↻ (2)
	After filling	↺ (1)	A or C (3)	↺ (3)				
	B. Hand lever pump	↻ (3)	C (2)	↻ (2)				
	After filling		A (1)					
Draining the truck	A. Motor Pump	↺ (4)	B (5)	↻	B (3)	↺ (1)	↻	↺ (2)
	After drainning	↻ (4)	A or C (3)	↻		↻ (1)	↻	↻ (2)
	B. Hand lever pump	↺ (1)	A (3)	↺ (3)				
	After draining	↻ (2)		↻ (1)				
Fueling the A-4	Motor pump and Disc indicator	↺ (4)	B (5)	↻	A (3)	↺ (1)	↺ (2)	↻
		↻ (4)		↻		↻ (2)	↻ (1)	↻
	After fueling		A or C (3)					

When fueling, watch the level indicator at the instrument compartment first and later, when the container is 2/3 full, the float level indicator (3). Prior to this, you must remove its screw cap.

Otherwise--you will be unable to see the indicator arm.

You can also obtain the fuel level from the measuring stick(4).
As soon as the container is filled you must shut down the B-Substance inlet.

2. Fueling through the drain plug

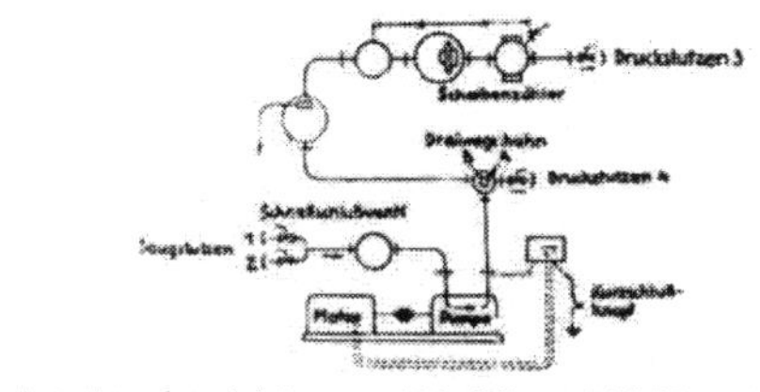

You may do this in two different ways.
You may fill the container by

A) utilizing the motor pump available
B) or the semi-rotary pump.

You will utilize the semi-rotary hand pump only if there is no motor pump available or if there is only a small amount of B-Substance to be pumped into the container.
You will find the instruments to be utilized when fueling through the drainage plug, in the instrument compartment.
The semi-rotary hand pump is stored there also. The motor pump apparatus is carried in a special pump trailer.

A. Fueling with the Motor pump

First you will prepare the motor pump apparatus for operation.

Unscrew the screw-caps of the suction-pipe connections (1) and (2) and the pump outlet (4).
Connect the suction hose to the suction pipes (1) and (2).
Screw the pressure-hose onto the pump outlet (4).
Open the cut-out governor valves on filler plugs (1) or (2) and on filler plug (4).
Turn the three stage faucet on position B, that is to the left.

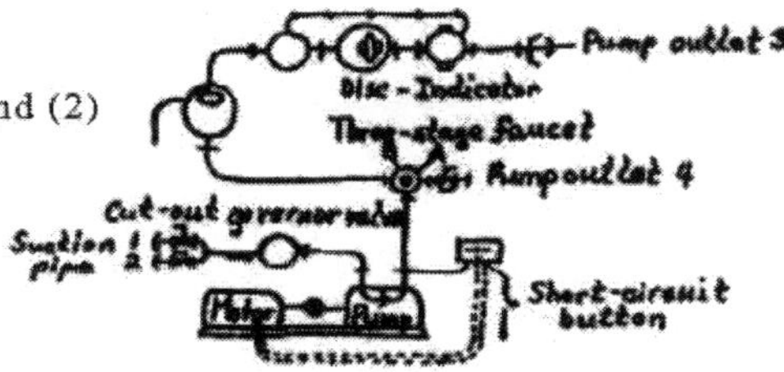

In the meantime you must perfor, the following duties in the instrument compartment:

(1) Connect the filler hose to the drainage plug.

(2) Turn the four-way faucet on position B.

(3) Close the cut-out governor valve.

Prior to starting up the pump motors check if the pump is filled with B-Substance.

Once the required amount of B-Substance has been pumped depress the short-circuit button on the motor pump. The motor will now stop. Close the cut-out governor valves on filler plug (1) or (2) and on filler plug (4).

After fueling you must turn the 4-way faucet in the instrument compartment on position A or C, that is to the left or to the right.

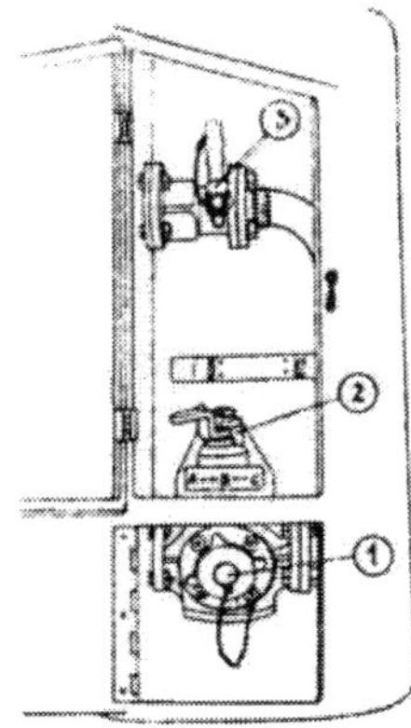

B. Fueling with semi-rotary hand pump

First you must screw the filler hose to the drain plug.

Turn the four-way faucet on Position C, that is to the right.

Open the cut-out governor valve.

Screw the pump handle to the semi-rotary hand pump.

Otherwise--you are unable to operate the pump.

Once you have finished with pumping, you must turn the four-way faucet on Position A, shut the cut-out governor valve and unscrew the pump handle.

DRAINING OFF THE KW CONTAINERS

When draining the KW container you will follow the same procedure you did ehen filling. Pay special attention to the position of the four-way faucet and cut-out governor valve. Drain only through the drainage plug.

A. <u>Removal by motor-pump</u>

Place four-way faucet on position B.

Close cut-out governor valve.

Start up motor pump. During removal watch the lever indicator in the instrument compartment and test the B-Substance with the measuring stick.

After draining place the four-way faucet on position A or C.

B. Removal by semi-rotary hand pump.
Place four-way faucet on position A, open cut-out governor valve.
After draining close the cut-out governor valve.

FUELING THE A-4

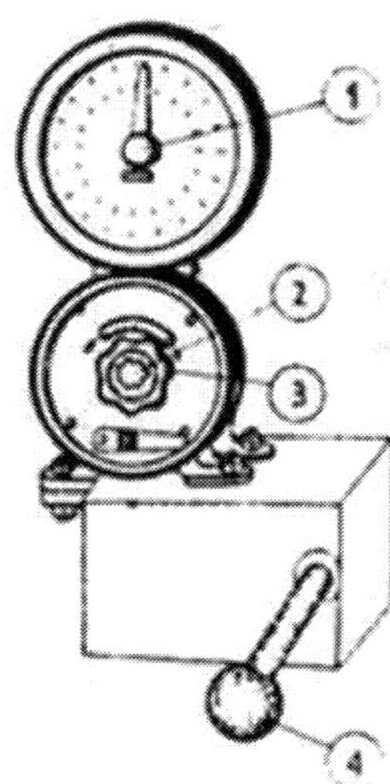

You will utilize the motor pump assembly. Place it somewhat behind the right middle wheel of the long range rocket carrier.
You need two B-Substance tankers for fueling, they are emptied at the same time.
Connect one suction line each to the drainage plugs of the tank-truck. The two other ends of the hose are to be connected to the two suction plugs (1) and (2) located on the pump-trailer.
Place the pressure line from the pressure plug (3) of the pump to the B-Substance ascending line at the long range rocket carrier.
Prepare the motor pump assembly for operation.

Open the cut-out governor valves at Plug (1), (2) and (3).
Turn the three-way faucet on position A, that is to the right.

You will measure the amount of B-Substance pumped with the disc indicated. Set the requested amount by adjusting the preadjustment device ofthe disc indicator. Divide the amount through ten. Then

(1) Make a rough adjustment by utilizing the large control button.
(2) Make a fine adjustment by utilizing the small control button.
(3) By utilizing the return button move all pointers of the recording mechanism on zero.
(4) Pull out the control rod. You hereby open the cut-off valve. Lock the rod.

In the meantime perform the following duties in the instrument compartment of the tank truck:
Place the four-way faucet on Position B.
Shut the cut-out governor valve.
Check if the pump is filled withB-Substance.
Then start up the motor.

Check if during operation of the pump small gas bubbles develop at the gas indicator night glass.
As soon as 1000 I of B-Substance has been filled, report it so that the B-Substance pilot valve is opened.

Once the set amount of B-Substance has flown through the pump it will turn off automatically.
Remember: Should the motor stop during pumping, then the amount set at the indicator is wrong.

After fueling

With the semi-rotary hand pump empty the ascending pipe and the hoses and unscrew the hoses.
Turn the four-way faucet in the instrument compartment on position A or C.

DRAINING THE A-4.

Connect the hose to the drainage plug of the B-Substance container and to the pipe connection with hand-operated valve at the ascending pipe.
Open hand-operated valve.

B-Substance will drain off automatically, but may however be sped up by utilizing the pump.

A-SUBSTANCE

From the supply depot to the launching site
the tank trailer is utilized to transport A-Substance. At the supply depot you will transfer the A-Substance from the railroad tank car to the tank trailer.

Prior to the transfer hould you detect dampness in the tank or any of its parts then you will dry the trailer with warm gas (nitrogen or air).
Otherwise--the valve will freeze up.

First remove all A-Substance or water from the A-Substance container.
Shut the warm gas line at filler plug (4).
Open all valves—that is, gate valve (5), exhaust valve (2) and (3), angle valve (9) and degasifier valve (10).

Remember: Valves mounted on lines with small diameter are to be opened wider than those with large diameter.

Otherwise--the warm gas will not be able to flow through all lines and instruments.

Unscrew the pipe connections at the manometer (11) and at the fluid level indicator (6) and turn them downward.

Otherwise--Water pockets will develop in the U-links.

Now let the warm gas run through. It lasts approximately six hours.
With a soldering lamp carefully pre-heat the knee-shaped pipe sections on the two safety valves (7) of the inside container and exhaust gas tubing.

Otherwise--water pockets will develop causing the safety valves not to function properly.

Once you have finished drying remove the warm gas line from the filler plug, shut all valves and reconnect the tubular connections to the manometer and liquid indicator.

FILLING THE A-SUBSTANCE TANK TRAILER

Remember: You will not connect all lines, especially not the suction lines. Do not forget the copper gaskets.

You will utilize the A-Substance pump-assembly. You will find it on the pump assembly trailer or in the railroad tank car.
Set up the pump assembly and secure it.

(1) Open the cut-off valve to the fluid level indicator.
(2) Open the cut-off valve to the degasifier tubing.
(3) Open the degasifier valve to the direct-degasifier-line.
(4) Connect the filler hose to the filler-cap and A-substance pump.
(5) Open the gate-valve by turning the valve handle.

POSITION OF THE VALVES FOR A-SUBSTANCE-FUELING AND DRAINING

Valve open   Valve closed
The number will give you the sequence.

Process		A-Substance-Transport Trailer						
		Filling line gate valve (5)	Exhaust Valve for: Direct exhaust gas line (3)	Exhaust Valve for: Exhaust gas cooling coil (2)	Shut-off valve for fluid level indicator (1)	Drain duct angle valve (9)	Vent valve (10)	Vent valve
Filling the trailer	Preparation	4	3	2	1			5
	Pumping							1
	After pumping	2	3					1
While driving and during prolonged parking of the trailer					1 Open only if necessary			
After the trailer has been drained completely				2	1			
After fueling the A-4	Preparation				1	3	4	2
	Pumping		3				2	1
	After fueling		3			1		2

Turn on the motor of the pump, but do not engage the pump as yet.

Open both degasifier valves of the pump and dry its interior by heating it with the warm-air apparatus. Open the cut-out valve of the suction line and the exhaust valve on the railroad tank car.

These valves may be moved at only a very little opening of the degasifier valves and then only until they are properly sub-cooled. Only then will you completely open the degasifier valves, screw them back one-fourth of a turn and let the A-Substance gas blow off.

Should A-Substance emerge from the degasifier lines, then turn the pump on and off several times. Do this first at short, and as you go along, longer intervals.

After approximately 5 to 8 minutes "engaging time" the manometer at the pump will jump to approximately 3 atu (atmospheric overpressure).

Then:

Leave the pump engaged and close both of its degasifying valves.

So: When filling the A-Substance into the container be sure not to go beyond the red marker on the fluid level indicator (6).

Once fueling is completed:

Disengage the pump and turn off the motor.
Furthermore shut the cut-out valve of the suction-line on the railroad tank car.
Open carefully the degasifier valve on the pump and shut gate-valve (5) on the A-Substance trailer.

Let the A-Substance evaborate in the hoses. This takes approximately 5 to 7 minutes.

Then loosen carefully the hose connections and remove the hoses.
Shut exhaust valve (3) of the direct exhaust line but leave cut-out valve (2) of the exhaust gas cooling coil open.
In conclusion test the safety valve (7) by slightly lifting and lowering the valve ball.

Parking and Driving the Loaded A-Substance trailer.

Cut-out valve (2) for exhaust gas cooling coil must be open.
Exhaust valve (3) for direct exhaust gas removal must be closed.

Watch the manometer. It should show little or no pressure at all.
The operating pressure amounts to 1.5 atu at the most.
From time to time remove the ice accumulated at the degasifier plug (8)

AFTER HAVING COMPLETELY EMPTIED THE A-SUBSTANCE TANK TRAILER

Shut the cut-out valve to the exhaust gas cooling coil.

Otherwise--dampness will enter the continer forcing you to redry it.

FUELING THE A-4

Place the A-Substance pump assembly into the holding device on the trailer and secure it.
Connect the line between suction plug of the pump and angle valve (9) of the trailer.
Connect the pressure line between the pressure plug of the pump and the A-Substance ascending line on the rocket carrier.
Open the cut-out valves (1) of the fluid level indicator.
Start up the motor but do not engage the pump as yet. Continue as you did above.

Remember: In order to pump you must open the angle valve (9) and the degasifier valve (10) of the gas trap.

Once the pump is finally engaged shut both degasifier valves of the pump and the degasifier valve (10) of the gas trap. Then open the exhaust valve (3) to the direct exhaust line.

Now We Start With Fueling the A-4

Watch the A-Substance degasifier-tube at the launching platform.
When the A-Substance stops running, it is an indication that the container is filled.
You must pay attention to many things,
in this particular case to gaskets and valves.

After Fueling

Disengage the pump and stop the motor.
Shut angle valve (9).
Carefully open the degasifier valves of the pump and shut the exhaust vale (3) of the direct exhaust line.
Connect the hand valve to the A-Substance fueling joint.
First loosen the hose connections.
Remove carefully the hoses and immediately shut the connection plugs.
Remove the A-Substance pump assembly and place it into the pump trailer.

Refueling
Connect the pump to the refueling device. Report that you are ready for refueling. Fill the container to the overflow.

Once you have finished refueling, disconnect the hose connection.

DRAINING THE A-4

Small amounts are drained through the refueling coupling.
Large amounts are drained faster through the fueling plug.

T-SUBSTANCE

Railtoad tank cars bring the T-Substance to the supply depot, sometimes the tank trucks are loaded on the railroad cars.

From the supply depot to the launching site

in any case, however, utilize the tank truck. In the instrument compartment of the tank truck you will find the pumps and the fueling, draining, and refueling device.

Open the doors at the end of the tank truck. You will then see the starting motor on the left, the motor pump in the middle, above it the semirotary hand pump, behind it the overpressure valve and the two reflex glass indicators, and finally the pipe lines with faucets and covers.

FILLING THE TANK TRUCK

You may fill the tank truck by utilizing the filler plug or the over-pressure valve. When filling through the filler plug you may use the semirotary hand pump or the motor pump. When filling through the over-pressure valve you may use the motor pump.
When doing this you have to set the three-way faucet (1) and the faucet (2) through (4) separately, operate the change-over valve of the semirotary hand pump and connect the hose different fittings.
Just what you have to do is shown on Page 118.

Before using the pumps
fill the water container of the tank truck.

Otherwise--you will not be able to wash off the areas contaminated by the T-Substance.

Position of the cock and valve when fueling with the T-substance

Stellung der Hähne und Ventile beim T-Stoff-Tanken

Arbeitsvorgang (Process)			Hahn 1 (Cock 1)	Hahn 2 (Cock 2)	Hahn 3 (Cock 3)	Hahn 4 (Cock 4)	Bemerkungen (Remarks)	Umschaltventil 5 d. Handfl.-Pumpe (Changeover hand pump lever 5)
Füllen (Filling)	über den Füllstutzen (Above fill sleeve)	mit Handflügelpumpe					Schlauchanschluß bei „A" und „B"	V
		mit Motorpumpe						D
	über Überdruckventil	mit Motorpumpe					Schlauchanschluß nur bei „A" (Hose connection at "A" only)	D
nach dem Füllen (After filling)								
Tanken (Fueling)		mit Handflügelpumpe (With hand pump)					Schlauchanschluß nur bei „B" (Hose connection at "B" only)	V
		mit Motorpumpe (With motor pump)						D
nach dem Tanken (After fueling)								
Umtanken (Transfer)		mit Handflügelpumpe					Schlauchausschluß bei „A" und „B"	V
		mit Motorpumpe						D
nach dem Umtanken (After transfer)								
Entleeren (Draining)							Schlauchanschluß bei Hahn „3"	

This water is also utilzed for rinsing the hoses as well as for washing off contaminated areas on the truck and on the tires.

Screw the glass flow indicators into the hose lines.

Otherwise--you will not be able to watch the flow of the T-Substance.

Connect both ends of the suction and pressure hoses first, then operate the faucets and valves according to the instructin chart.

Otherwise--you might be endangered by the T-Substance remainder in the pump.

Connect all hoses and lines well.

Otherwise--the flow of the T-Substance will be very poor and its leakage create additional hazards.

Open the lid to the container.

Otherwise—the pump will not operate correctly at filling and draining. The air filter must be open.

At the Motor Pump

Fill up the two side containers of the frame with water; The water will thin out the T-Substance coming through the sealing ring.

Otherwise--The T-Substance would be dangerous.

From time to time empty those vats and refill them with fresh water. Fill the rotary pump with T-Substance.

Otherwise--there will be no suction.

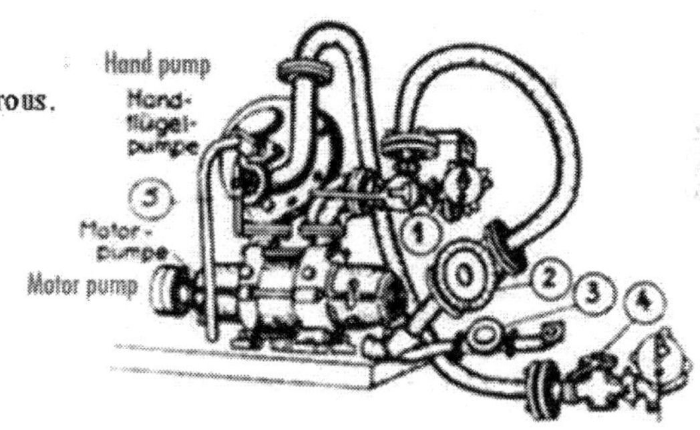

After Filling

Turn the three-way faucet (1) to the left.
Close faucet (4).

Otherwise-Due to the pressure of the gases T-Substance will spray from the pump-housing.

FUEL TRANSFER

Should you be required to transfer T-Substance into your tank truck then utilize the chart on page 1 1 8 for adjusting the faucets and valves.

Draining the Tank Truck

Park your vehicle so that it will slope to the back. Follow instructions on the chart. You will not utilize any pumps for draining.

FUELING THE A-4

In order to fuel the A-4, fill the T-Substance measuring tank located near the erecting-frame of the Rocket carrier first.
For filling it you may utilize the rotary hand pump or the motor pump.
The position of the faucet and valves is shown on page 118 under the heading "Fueling".

Be careful: The T-Substance measuring tank is filled by the motor pump in less than one minute.
Fill the measuring tank to the overflow.
Your comrad from the propulsion-unit crew will notify you as soon as the container is filled.
He will then empty it into the A-4 and let the contents of the overflow container run into the tank truck.
After fueling pump the hoses dry and flush them with water.

DRAINING THE A-4

is done by utilizing the drain cock.
After draining flush the container thoroughly.

Remember: The fire truck carries water to put out T-Substance fires.
Put out T-Substance fires only with water, never use the tetra-fire extinguisher.
Use only water for fighting T-Substance fires.

Z-SUBSTANCE

Z-Substance is transported in cans.
Be careful: Never transport T- and Z-Substance on the same truck.
Never fill them in the same container, not even if you have rinsed the container well.
Transport T- and Z-Substance separately.
Otherwise you will end up in a hospital.

REFUELING THE A-4

Your comrad from the propulsion-unit crew will fuel the A-4 with Z-Substance.

DRAINING THE A-4

Open the drain cock
Remove the contents
Afterwards flush the container and pipe lines thoroughly with water.

MORAL: Should living be worth anything to you
Then never smoke when fueling.

IN THE REMOTE CONTROL POSITION

A firing site contains two remote control sites: the Combustion-Cut-off and Guide Beam position. Both are installed behind the Firing site.

The radio installation is located at the combustion cut-off site where upon s certain velocity the power plant of the A-4 is cut off.

Whether or not the A-4 will reach the required firing range and destination depends upon how well you have done your job.

The guide beam position is set up only for launching requiring utmost target accuracy. With your guide beam devices you are able to loosen the deflection of the A-4 considerably.

It often depends upon the right moment
at which you should cut off the light.

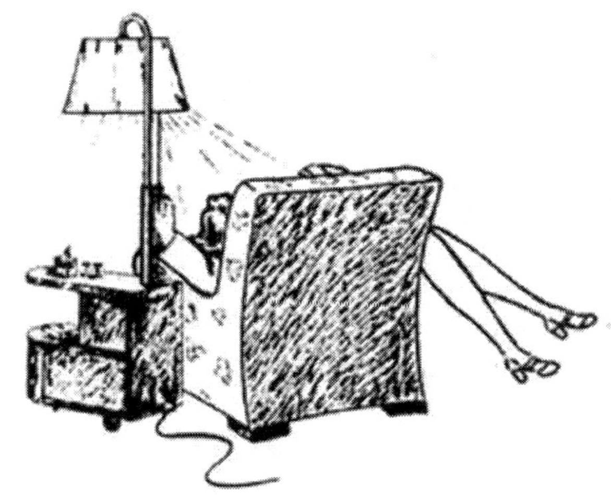

The power plant in the A-4 must be cut off at the right moment. It is called combustion cut-off.

Should combustion cut-off be given at a high velocity of the A-4, it will reach a long distance. If combustion cut-off is given at a low velocity of the A-4, it will reach only a short distance. It therefore is important to give combustion cut-off at a very certain velocity of the A-4.

Only then will the correct firing range be attained. The following belongs to the combustio cut-off ground installation.

1. <u>The Transmitter Truck</u>

   mainly contains all the different transmitters.

2. <u>The Cut-off Evaluation Truck</u>

   contains the V_0 test-receiver and frequency measurement device.

3. <u>The Dipol-trailer,</u>

   carries the dipol antenna with accessories and the high voltage cable for the cutoff signal transmitter.

4. <u>The Generator Trailer</u>

   produces the current for operating your installation.

You may divide your job at the combustion cut-off installation into the following catagories:

1. Preliminary duties—this includes mainly the erection of the antennas and cabling.
2. Duties in the transmitter truck.
3. Duties in the cut-off evaluation truck.

PRELIMINARY DUTIES

After receiving the order from the officer in charge, park the vehicle in the areas designated by the surveying crew.

Erection of the Rhombus-Antenna for the V_a-test-transmitter.

You will find the antenna in the transmitter truck. For erection utilize a measuring tape. It is divided into four equally long pieces and one longer piece by knots.

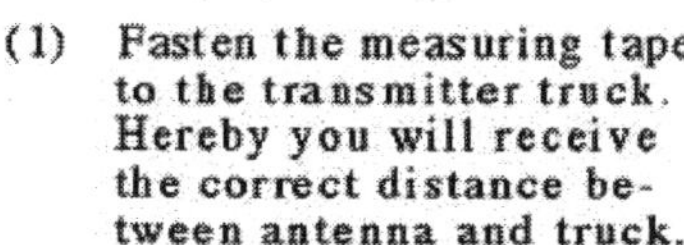

(1) Fasten the measuring tape to the transmitter truck. Hereby you will receive the correct distance between antenna and truck.

(2) Place the long piece in the direction of fire.

(2) Stretch the four short pieces rhombically. The four knots will designate the corners.

(4) Drive the mast supports in at the corners.

(5) Place the antenna rods into the mast supports. The antenna rods are assembled from two parts.

(6) Bend the antenna rods toward the inside.

(7) Lay out the antenna wire and connect it.

(8) Place the guy ropes on the hook of the antenna rod. Straighten the antenna rods and anchor the guy-ropes.

Important: Always connect the antenna wires to the vehicle immediately.

ERECTION OF THE RHOMBUS-ANTENNA FOR THE Vo-TEST-RECEIVER

You will locate the antenna in the cut-off evaluation truck where it will be set up in the same manner you Set up the antenna for the Vo-test transmitter.

Remember however: the measurement are half as large. Each rod is anchored by only one guy rope.

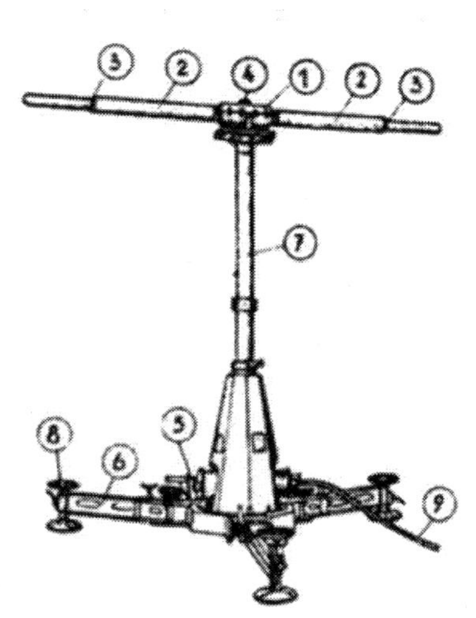

Erection of the dipols for the cut-off signal-transmitter

Park the Dipol trailer in line with the Rhombus-Antenna. The Distance from the cut.off evaluation truck is approximately 25 m.

(1) Place the Dipol head on the folded mast.

(2) Insert the Dipol arms.

(3) Adjust the arms lengths according to the calibration chart.

(4) Adjust the condenser Ci according to the calibration chart.

(5) Adjust condenser C2 according to the calibration chart.

(6) Unfold the footings.

(7) Erect the Dipol upright in the direction of fire.

(8) Disconnect the chassis. Solder the Dipol diagonally to the ground spindles.

(9) Reel off the antenna cable and lay it out. Lay out no curves with less than 1 meter in diameter.
Camoflage the chassis.

Making the Connections

Lay out 3 Field-Trunk Cables and 1 High Voltage Cable from the cut-off evaluation truck to the transmitter truck.
Lay out a cable from the current supply trailer to the cutoff evaluation truck or transmitter truck.
Connect the cables on the outside of the truck; the high voltage cable is connected on the inside of the ruck.
Remember: Cable couplings are designated by color and are not to be mixed up.
Drive one ground conduction pipe each into the ground at the cutoff evaluation truck, the transmitter truck and the current supply trailer and connect it to the ground terminal of the truck.

MORAL: Your work will progress well
Should all preparations have been made properly.

Besides other devices:

The transmitter truck contains the V_0-Test-Transmitter; transmit a certain number of electrical waves per second from which the speed of the A-4 is calculated.
The normal frequency auxiliary transmitter transmits exactly twice as many waves as the V_0-Test-Transmitter. These waves are utilized for the Vo-Test-Transmitter.

It further contains:

The cut-off signal transmitter; it gives the signal for combustion cut-off.
The phantom signal modulator; it gives a disguise for the combustion cut-off signal.
The cut-off signal modulator; it gives the correct combustion cut-off signal.
The cut-off signal control receiver; it is similar to a cash register. It checks to see if the signal given is correct and registers same.

The individual devices must be:

adjusted,
engaged,
turned, according to the ordered frequency.

Prior to each launching you will receive an order informing you of the frequency to be used. It might, for excample, be

green 3; ABC-425; D-1; E-2--O; 56Sec.; red 6; 301B; Abel; Long Transmission.

Remember: "green" always refers to the V_o-Apparatus.

"red" always refers to the control unit.

How to adjust the individual devices will be mentioned in detail.

Starting Up the Transmitter Truck.

Give the order to start up the generator set.
Depress the switch buttons (1) and (4), through (5) engage voltmeter (6) and test through switches R, S and T. The deflection of the pointer must be 220V.
Through (7) engage the control panel and through (8) the battery voltage.

Place lever switch (9) in a downward position and test take-off contact and contact timer.

As a result lamp (10) contact timer will go out.
After 50 seconds you will hear the antenna contactor operate. After an additional 10 seconds it will fade out.

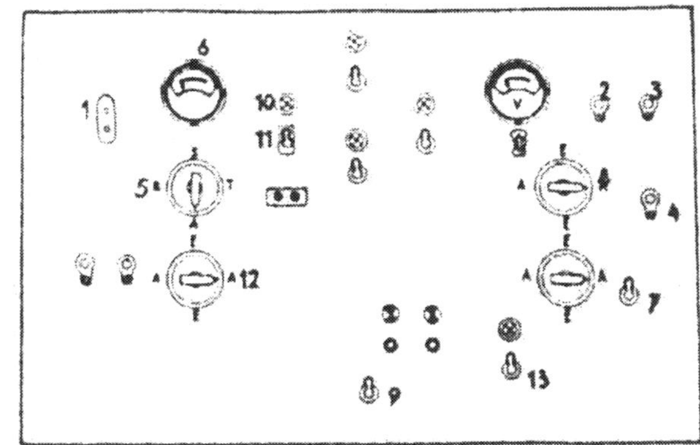

Place toggle switch (11) in a downward position thus the contact timer will go into the zero position.

When lamp (10) produces a dark flicker, the return movement has ended.

Now place toggle switch (11) back into the neutral position.

Light (10) burns with normal brightness.

Turn switch (12) on "on".

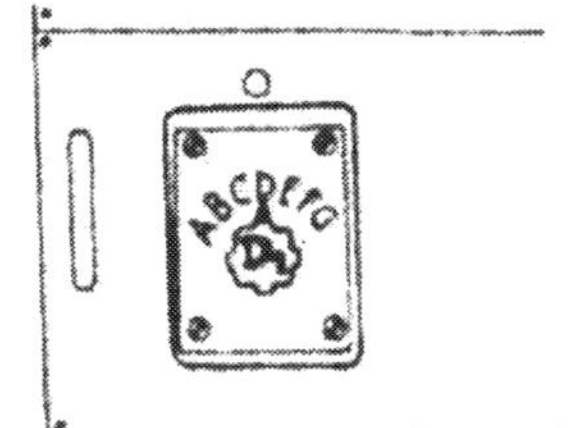

A. THE CUT-OFF SIGNAL TRANSMITTER

In addition to the power supply unit the cut-off signal transmitter contains four fields:

Field I	Control Panel
Field II	Control and Amplifier Stage
Field III	Amplifier and Modulator Stage
Field IV	Final Power Stage

<u>Adjusting according to the ordered frequency</u>

The ordered frequency contained the following specifications: red 6; 56 Sec., Long Transmission

<u>Adjusting the Carrier Frequency red 6</u>

Open Shutter 1 in Field II.

You will find the rotary switch D 1 behind the shutter, adjust it according to the frequency. For example, for red #6 you will turn it on D.

Close shutter 1 again.

Setting the Long Transmission and Time

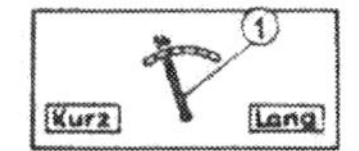

In the Relay Box: Place the short circuit plug on "long".
Turn the Selector Switch (1) on 56 Seconds

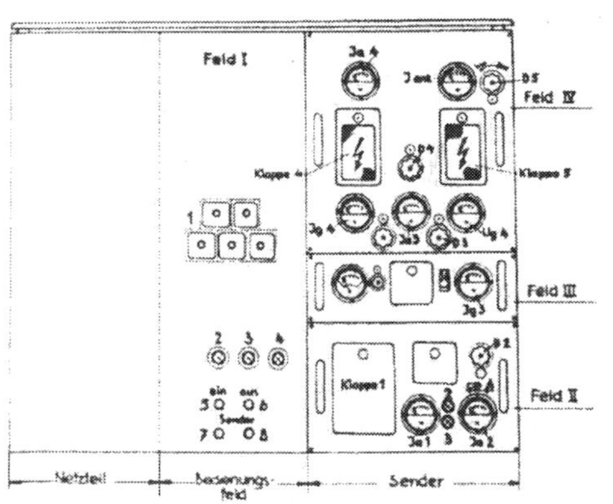

Engaging the Transmitter

In Field III turn the Tumbler Switch to the top.

Then--you will have better synchronization.

In Field IV turn Button D 5 on "fixed".

Then--the tube outputobtained will be used up by the antennna short circuit (electrical bulb).

Otherwise--the anode tins will glow while and the tube will be destroyed.

In Field I Depress all automatic cut-outs (1).

In Field II Bulb (2) and (3) must light up.
Lamp (2) or (3) will go off as soon as the correct temperature for the thermostat has been reached.

In Field I Depress Button (5) from 2 to 3 Seconds.
Then--a motor with switch mechanism will start up.
The red signal lamp (2) must light up.

Otherwise—the filamentvoltage grid and screen grid voltage is not turned on.

Remember: Should lamp (2) not light up immediately then depress Button (5) up to 3 times.

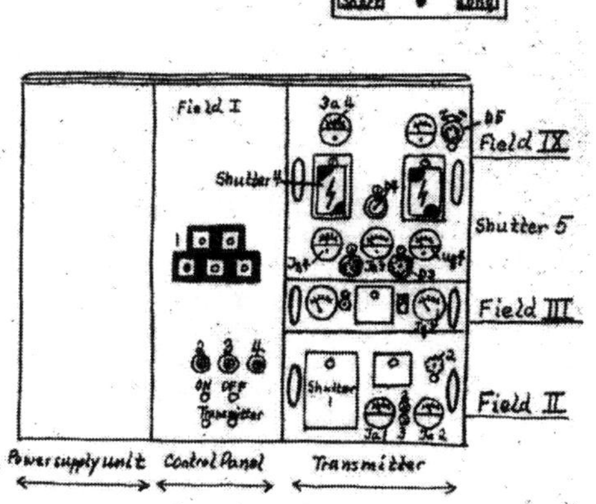

In Field II the deflection of pointer Ja 1 must amount to approximately 15 mA.
Important: After every trip made by the transmitter truck let the transmitter operate for 15 minutes. Before you make any further connections:

In Field I Depress Button (5) from 2 to 3 seconds.
After approximately 3 minutes signal lamp (3) must light up. This will tell you that 800 volts are turned on.

In Field IV the deflection of the pointer Ug 4 must amount to approximately 80 to 100 scale sectors. In

Field I Depress Button (5) from 2 to 3 seconds.
After approximately one minute the red signal lamp (4) must light up.
It will tell you that 1500 volts are turned on.

Synchronizing the Transmitter

In Field II Turn lever switch (4) to the right on "Synchronization".
By utilizing Button D-2 regulate the deflection of the pointer Ja 2 on minimum value and Jg 3 in Field III on maximum value.

In Field IV Set the deflection of the pointer Ja 3 on minimum value by utilzing Button D 3.
In doing this the deflection of the pointer Jg 4 must show a maximum value.
Turn Button D 4 until Ja 4 shows the smallest deflection.
At the same time Jant must show the highest deflection.
Turn Button D 5 to the left on loose and again synchronize with D 4.
Synchronize the measuring device Jant on maximum value by slowly turning the Button D 5 to the right.
The electrical bubs of the mute antenna will then light up the brightest.

You will now again synchronize the transmitter by utilizing Buttons D 1 to D 5.

In Field III After completing synchronization turn lever switch in a downward position (center Line) so that the deflection of the pointer of the measuring device Jant in Field IV will approximately drop in half.

In Field II turn lever switch (4) to the left on "Operate".

After making these adjustments report that the cut-off signal transmitter is ready for operation.

Important: In Field IV the anode tins of the power tubes behind shutter (4) and (5) may produce a dark glow only.

Disconnecting the Transmitter

In Field I you have the push-buttons (6) and (8) for disconnecting the transmitter. By utilizing push button (8) you will only disconnect the 1500 v stage. You will utilize it should the transmitter not be used for the time being, but must be ready for operation again in a very short time. By utilizing push button (6) all voltage is off should you desire to turn off the transmitter completely.

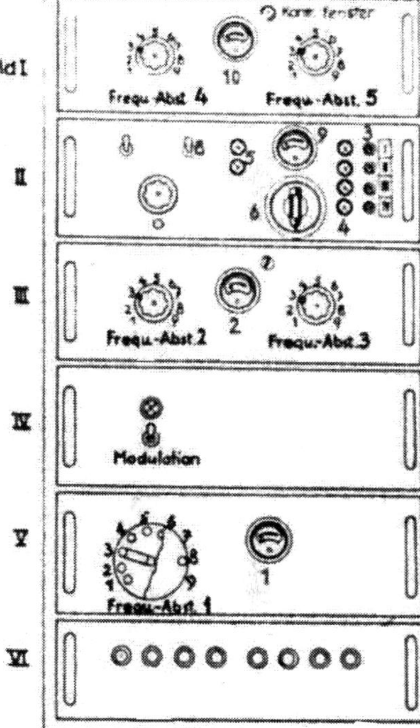

In order to operate the V_0-Calibrating-Transmitter you must make co at the first cabinet as well as at the output cabinet..

Adjustment According to Ordered Frequency

The ordered frequency requested: green 3.

Adjusting the Frequency green 3

First cabinet:

In Field V — Remove the cover of the "frequency adjustment device 1". Turn the adjustment on position 3.

In Field I and III — Set the buttons of the "frequency adjustment device 2 to 5" on the ordered frequency value 3.

Remember: In Field II all switches must be turned off.

Output Cabinet:

Close or open the two latchets (1) for the coupling spools (2) on the output cabinet as shown on the table located above frequency "green 3".

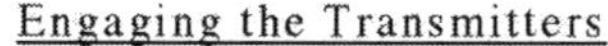
Engaging the Transmitters

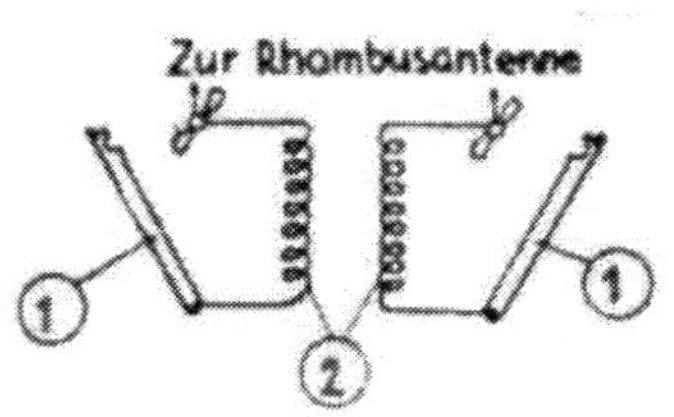

Instruction Table for the Transmitter Truck.

Engage Switch (13).

Output Cabinet: Place both slide latchets (3) toward the front on "mock antenna".--In Field I depress the five automatic cutouts (1).

Depress pushbutton (9); this must draw the antenna contactor in the output cabinet audibly.

The ventilator contactor of the output cabinet may be watched through the back right truck door.

Remember: The voltage of the transmitters can be engaged only in Field I of the output cabinet.

In Field II turn knob (1) on "stable".

In Field I depress Button (5) from 2 to 3 seconds. A motor with switch mechanism will then start to operate; after approximately 3 seconds a green signal lamp (2) will light up.

Otherwise--filament voltage and grid voltage are not turned on.

Remember: Should lamp (2) not light up then the switch arrangement was not in the correct position normally. Repeat the engaging of Button (5) up to 3 times.

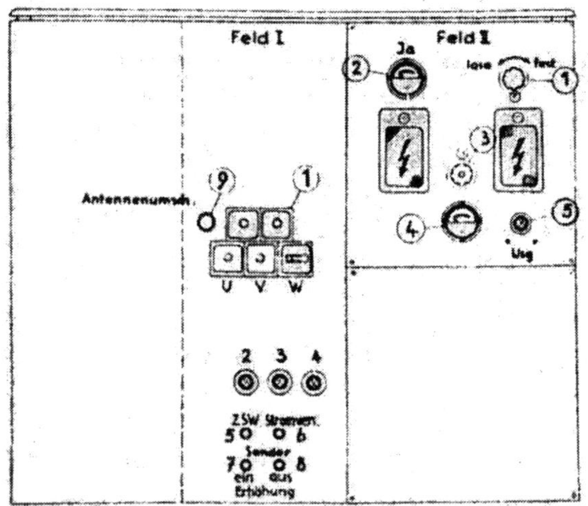

First Cabinet

In Field II the control lamps (3) must light up.

Important: After each trip made by the transmitter truck or after exchanging one of the rectifier tubes you must let the transmitter operate for 15 minutes. Then do not make any further connections.

Output Cabinet

In Field I depress Button (5) again from 2 to 3 seconds. The switching motor will start up again. After 2 minutes the signal lamp (3) must light Up green. It will tell you that an 800 volt current has been turned on.

In Field II the measuring device (4) will show a deflection.

In Field I Again depress Push Button (5) 2 to 3 seconds.
The switching motor will start up again. After approximately 3 Seconds signal lamp (4) will light up. This tells you that 1500 volts are turned on.

Remember: First Cabinet: In Field II the four control lamps (4) and the two control lamps (5) must light up. In Field V the deflection of the pointer (1) must be located in the green zone.

Synchronization of the V_0-Calibrating-Transmitters

First Cabinet:

In Field II turn the measurement selector switch (6) on Position 2—III and depress.
Watch measuring device (9).

In Field III turn the frequency adjustment Knob (2) until the measurement device (9) in Field II shows the largest deflection. By utilizing the turn knob of the "frequency adjustment device *3*" the Measuring Device (2) is synchronized with the highest value shown.
Remember: The Knob "frequency adjustment 2" and "frequency adjustment 3" buttons are not required to be accurately set on the previously adjusted number any more*

Output Cabinet:

In Field II adjust the turn Knob (3) in such a way that the brightness of the electric bulbs (mock antenna) are at maximum.
The deflection of the pointer of the measuring device (2) returns to the lowest value.
Retune it by utilizing push button (1).

Remember: The output cabinet must be turned simultaneously with the first cabinet.

Engaging and Returning the Normal Frequency Auxiliary Transmitter

The Vo-Calibrating-Transmitter must be engaged and turned prior to starting up the transmitter.

Connections are made only in the first cabinet

In Field II Place the lever-switch (8) into the upward position.
Turn the measurement selector switch (6) on position 2—1 and depress.
Watch measurement device (9).

In Field I Turn the Frequency Tuner Knob 4" until the measurement device (9) shown its largest deflection. Tune the measuring device (10) with the highest indicator value by utilizing "Frequency tuner 4".
By looking through thecontrol window check if the electric lamp has maximum brightness.
Remember: Adjustment of the turn knob "Frequency tuner 4" and "Frequency tuner 5" is not required to coincide with the number set when the sender was switched on.

At the Service Panel of the Transmitter Truck

Turn Switch (13) on "off".
By doing this the electric bulbs of the mock antenna will go off.
Now make the report that the Vo-Calibrating Transmitter and the Normal Frequency Auxiliary Transmitter are ready for operation.

Disconnecting the Vn-Calibrating Transmitter

Differentiate:

1- Case: Temporary disconnection {up to a few hours).
2. Case: Final disconnection.

1. Case: Should you depress Push Button (8) in Field I of the output cabinet, then the anode current of 1500 volts is reduced to 1300 volts.
Should you depress Push Button (7) in Field I then the anode current will rise to 1500 visits again.

2. Case: Should you depress Push Button (6) in Field I of the output cabinet then all current is removed from the transmitter. All control lamps will go out.

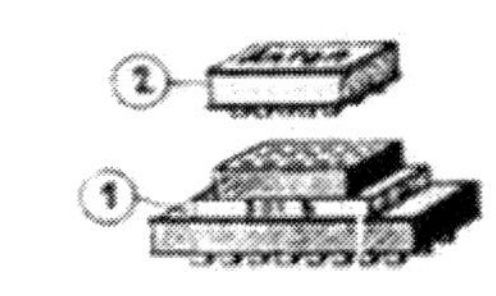

C. THE CUT-OFF SIGNAL MODULATOR

Setting the ordered frequency
for example, 301 B; Anton; red 6,

(1) Place the Frequency Selector-Plug 301 B on the counterpart.

(2) Place the sequency selector plug "Anton" on the Frequency Selector-Plug.

(3) Depress the automatic cut-out.

(4) Engage the Switch at the cut-off signal-modulator.

D. THE PHANTOM SIGNAL MODULATOR

Setting the ordered Frequency, for example, 301 B.

(1) Plug in the Phantom-Signal Modulator key B.

(2) Engage both automatic cutouts.

(3) Watch if the white control lamp at the Phantom-Signal Modulator lights up.

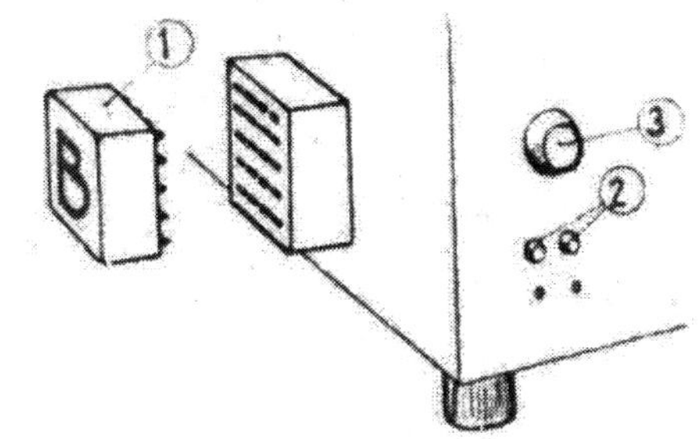

E. CUT-OFF TEST RECEIVER WITH TRANSFORMER

Setting the ordered Frequency, for example, red 6: Anton.
Plug in the high frequency insertion Number 6.
Plug in Sequence Selector plug "Anton".

In order to test the cut-off signal modulation the cut-off signal transmitter must be engaged and turned.

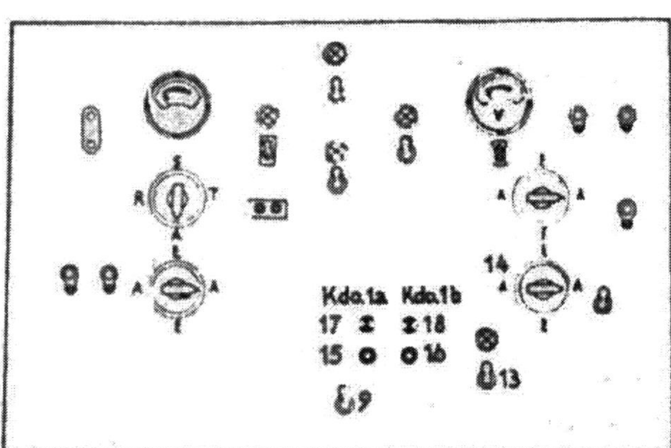

At the Service Panel of the Transmitter Truck
Turn Switch (13) on "in".
By doing this the Transformer will start up.
Simultaneously depress the Button (14) and (15).
Watch if Indicator (16) signal 1a and after
3 Seconds Indicator (17) Signal 1b will appear.

Testing with Cut-off Evaluation Truck

After receiving the order from the cut-off evaluation:
"Test with V_0-Calibrating-Transmitter and engage Modulation".
At the service Panel of the Transmitter Truck:
Move Switch (13) on "in".
(The V_o-Calibrating-Transmitter must be engaged.

First Cabinet:

In Field IV on the Switch lever "Modulation".
The lamp located above it must light up.

After completing the test:
Place Switch (13) and Switch Lever "Modulation" again on "off".
After receiving the order: "Turn on, all devices for signal testing" from the cut-off evaluation'truck, you will follow the proceedings of the previous test.
In addition engage:

the Cut-off Signal Transmitter,
the Cut-off Signal Modulator,
the Phantom Signal Modulator and
the Transformer.

Call up the Cut-off Evaluation Truck and report: "Transmitter Truck prepared for signal testing."
After receiving the order from the Tansmitter Truck: "Test the take off".
At the Service Panel of the Transmitter Truck place:
Switch (9) "take-off test" on "in". The contact timer will now start up.

The indicators (17) and (18) are actuated from the cut-off evaluation truck.
Report the appearance of indicator signal 1a (17) and signal 1b (18) individually to the cut-off evaluation truck.
Turn Switch (13) at the Service panel of the Transmitter Truck and Lever Switch "Modulation" at the first cabinet on "off" again.

Preparing the V_0-Calibrating Transmitter for Launching

My friend, you must now be careful:
Prior to toughing the lash you must
Disconnect the Automatic Cut-out..
Otherwise you will be carried out as a corpse.

Thus:
1. Move: In Field I turn off automatic cut-out W at the output cabinet.
2. Move: At the output cabinet push both lashes (3) back on "Rhombus Antenna".
3. Move: At the output cabinet engage automatic cut-out W in Field I.

After receiving the order; from the Cut-off Evaluation Truck: "Transmit the Vo-Calibrating-Transmitter for 1 Second".
At the output cabinet: In Field I depress Button (9) for 1 Second.

Be Careful! Your Antennas Glitter".

MORAL: Depress the Button for only 1 Second,
so that the enemy cannot detect you.

Here it counts to determine
MOTTO: Velocity through electric waves,
So that one may be able to determine
The correct time for combustion cut-off.

The speed of the runner is determined by utilizing a Stopwatch.
The velocity of the A-4 is determined by utilizing the cut-off Evaluation Truck.
The Cut-Off Evaluation Truck contains primarily:

the Vo-Calibrating Transmitter; it receives electric waves which are transmitted by the Frequency Doubter of the A-4. Per Second there are less waves transmitted than from that of the normal Frequency Auxiliary Transmitter. The difference is a measurement for determining the Velocity of the A-4.

The Frequency Device: At cut-off velocity of the A-4 it signals the Cut-off Signal Modulator making the latter form the combustion cut-off signal and relay it to the cut-off signal transmitter.

All instruments of the Cut-off Evaluation Truck are stored in the Vo-Calibrating Transmitter. It contains 9 Fields and looks like this:

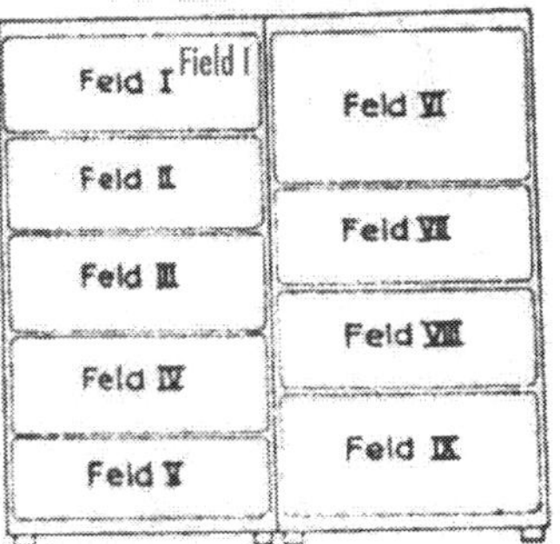

Field I	Control Pane! and Inspection Panel
Field II	Bridge Amplifier
Field III	Resonance Amplifier
Field IV	Vo-Calibrating Transmitter
Field V	Empty
Field VI	Frequency-Resistance Bridge
Field VII	Empty
Field VIII	Audio-Frequency Oscillator
Field IX	Main Connected Unit

You are to engage the V_0-Measuring-Device
and test its tubes, adjust it
according to the ordered frequency,
test and
ready for launching.

Putting Into Operation

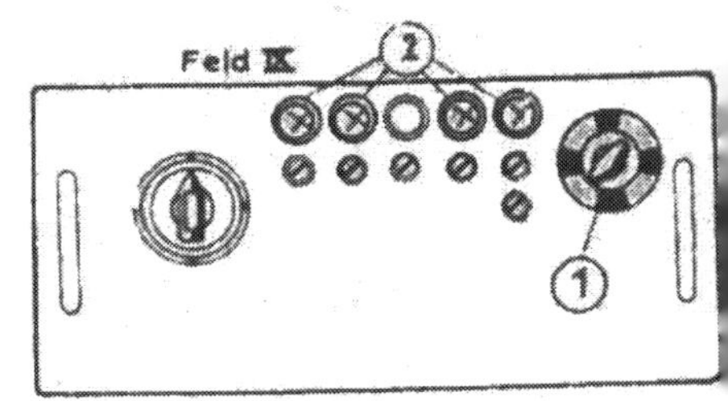

In Field II Turn on Switch (1).
Check, if after one Minute the Control-lamp (2) -not the one in the middle-lights up.
Remember: The Current consisting of 220 V/50 Hz may fluctuate between 210 and 230 V only.

Otherwise--the entire measuring device will not operate correctly,

Should one control Lamp go out, then disconnect immediatley and rectify the fault,

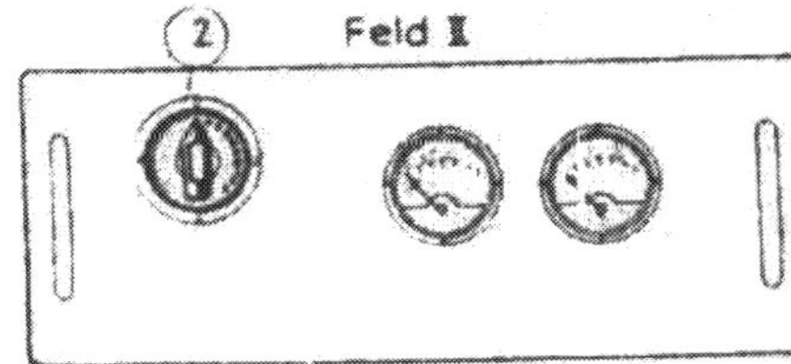

Testing the Tubes

utilize in Field IV the Measureing Device (1).
In Field II depress Selector Switch (2) in position 1, 2, 3, 4 and 6, The-deflection of the pointer of the Measuring Device (1) in Field IV must be located in the red zone.

Remember: Position 5 of the Selector Switch is designated by a ring, A tubular test in this position is possible only then when testing without the transmitter truck has been completed.
You will find tubes 1 to 4 in Field II;
Tubes 5 and 6 in Field I,

In Field III Depress Selector Switch in Position 1, 2, and 5, 6, 7. The deflection of the pointer of the Measuring Device (1) in Field IV must be in the red zone.

In Field I Place the Switch marked "Testing" on the middle position.

In Field III Depress the Selector Switch in Position 3 and 4 (designated by a ring). The deflection of the pointer of the measuring device (1) in Field 4 must be located in the green zone (0,05 - o,2 mA).

In Field I Turn the Switch marked "Testing" to the right.
In Field VIII Depress the Selector Switch in position 1 and 2. The deflection of the pointer of the measuring device (1) in Field IV must be within the red zone.

In Field IV Turn Button T to the right until the Number 1 appears at the circular opening above the control Button T. Depress Selector Switch (2) in Position 1 to 5,

The deflection of the pointer of the measuring device (1) must be within the red zone.

Turn Control Button T to the left until Number 100 appears at the circular opening.

Depress Selector Switch (2) in Position 7 and 8; the deflection of the pointer (1) must be within the red zone.

Remember: Should you not get into the required zones, then exchange the tube in question,

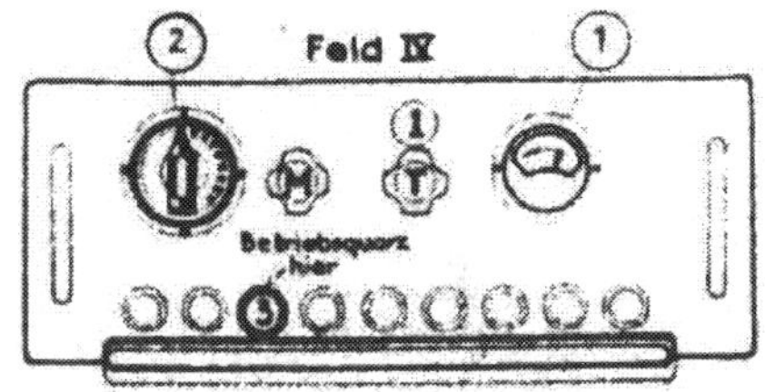

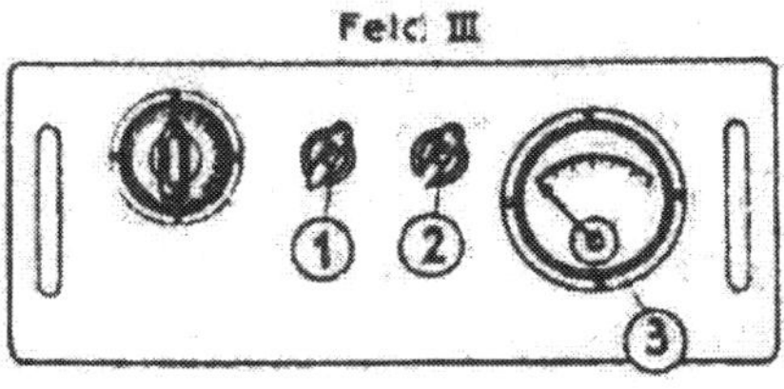

Setting Accordjng to the Ordered Frequency

The following Frequency Specifications were ordered:,
E = 3-0;ABC = 425; D = 1; green 3.

Setting the Resonance "3 - 0"

In Field 111 Turn Switch (1) on Position "3".
Switch (2) on Position "0",

Setting the Oscilator Stage "green 3"

In Field IV Open the shutter and place the Crystal "Po 3" into the two white-edged sockets designated "crystal". Then close the Shutter again.

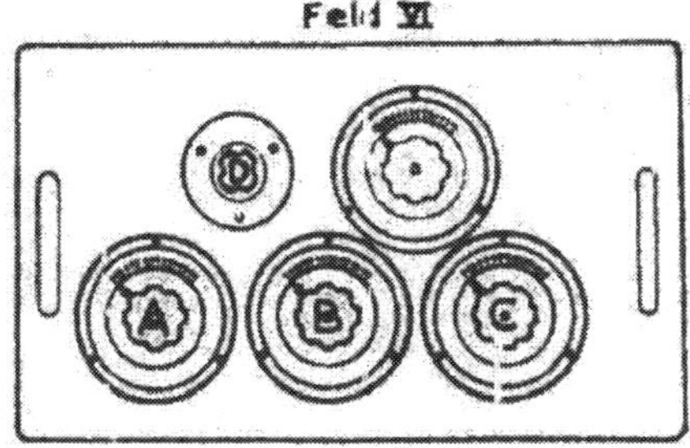

Setting the Launching Frequency.."425"

In Field VI Turn the Selector Switch A on "400".
Selector Switch B on "20"
Selector Switch C on "5"
Control Button D on "1"

Remember: The Selector Switches must be correctly engaged.

THE THREE TESTS OF THE V_0-MEASURING-DEVICE

Prior to putting on a show the Bicycle Acrobat will perform three tests. They are:

first, she will check the bicycle,
second the balancing act;
third, if her movements coincide with the music.

You will also conduct three tests in regard to the V_0-Measuring-Device:
first, you will test the instrument of the measurement device by itself, that is, without the Transmitter,
secondly, you will test the measurement device together with the Vo-Calibrating Transmitter in the Transmitter Truck.
third, you will test the collaboration of all devices of the BS ground installation, including the cut-off Signal Transmitter.

Now pay attention:

1. Case: Testing without Transmitter Truck

Even thorough you are only testing the instruments of the Indicator Device without the Transmitter, you must notify the crew of the Transmitter Truck, since the connecting cables are already payed out. Otherwise--testing in the Transmitter Truck may cause interference.

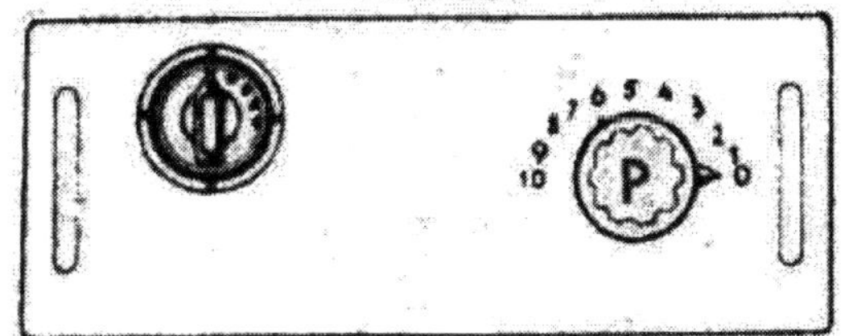

A. Testing the Frequency Measuring Device

In Field VIII Place Regulator Button P on Position 0.

In Field IV Turn the Button marked T to the left until the number "100" appears, in the circular opening.

In Field ! Place the lever-Switch "Testing" to the right. By doing this the control lamps (1) and "Transmitter" must light up.

In Field VIII Turn Button P slowly to the left

The following Pointer Deflections and Control Lamps must appear now:

a) In Field II the Deflection of Pointer (3) will slowly increase and the deflection of pointer (4) will from the zero position to the left

b) In Field II the Measuring Device (3) has reached the green zone 0,2 to 0,6 mA.

In Field II the Lamp "Reception" will light up.

c) In Field II the pointer of (4) will slowly return to zero; In the meantime the pointer of (3) is within the red Zone.

In Field I at the Zero Position of the Measuring Device (4) the lamp "Warning Signal" will light up.

d) In Field II after returning to the Zero Position the pointer of (4) will immediately return to the left again, due to the fact that the resistance bridge is now set automatically by a relay on the main cut-off signal Frequency.

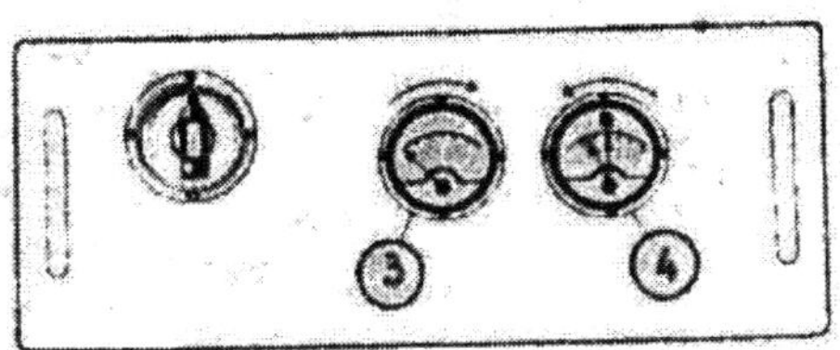

Remember Control Button P must always be slowly turned to the left in Field 8 causing the thusly formed Tone in Field I to be heard in Loudspeaker (2). The Measurement Device (3) in Field III indicates nothing as yet.

e) In Field II the pointer of (4) goes through the Zero Position for the second time.

In Field I the Lamp "Home Signal" will light up, at the same time the Lamp "Warning Signal" will go out.

d) In Field II the Pointer of (2) will go back. At a deflection of 0,2 and 0,1 mA the following will die out:

In Field I Control Lamp "Reception". Lamp "Home Signal" remains lit.

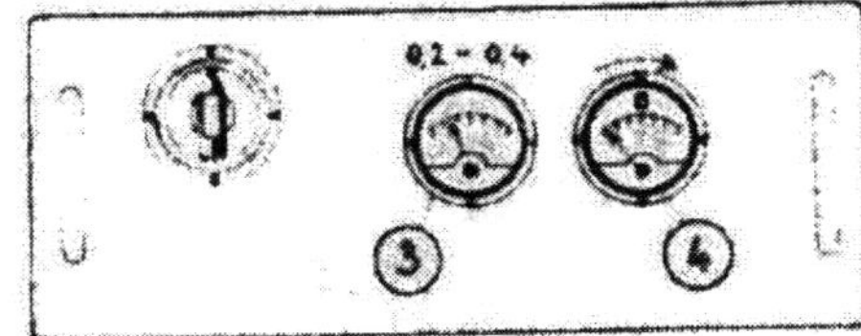

Remember: In order to repeat the test, Control Button P in Field VIII must be returned to the zero position. The Lever Switch "Testing" in Field I must be put on the middle Position for a short time and then returned to the Position on the right.

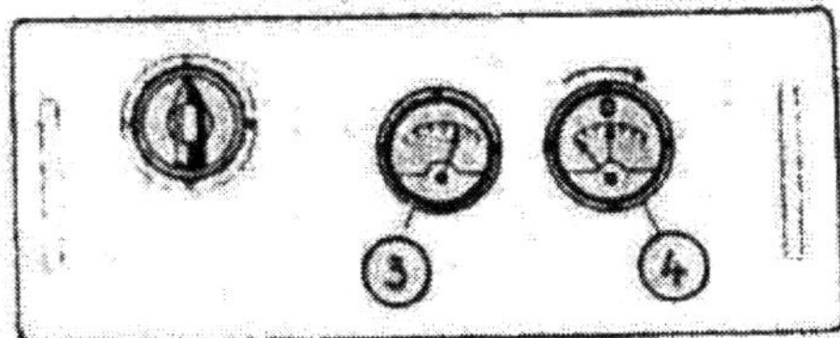

B. <u>Testing of the Signal Release by Manually-Given Cut-off Signal</u>

This test takes place in Field I only.
Depress Button "Take-off", then the signal Lamp "Take-off must go out.
Place Lever Switch "Testing" to right:
Lamp "Transmitting" will light up.

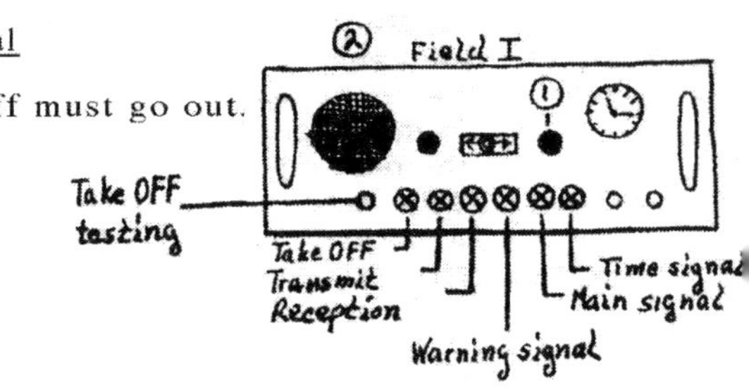

Depress Button "HS Hand BS" several times.
No Signal Lamp should now light up.

Depress Button "VS Hand BS".
Lamp "Warning Signal" must now light up.

Depress this button again;
Nothing should change.

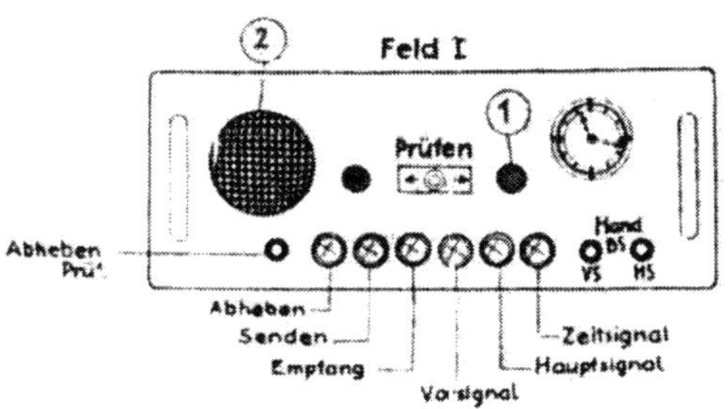

Depress Button "HS Hand BS".
Lamp "Warning Signal" must go out, and at the same time
Lamp "Home Signal" and "Time Signal" must light up.

Depress this Button again:
Nothing may change.

2. Case: <u>Testing with the Vo-Calibrating Transmitter</u>

In order to conduct this test a high voltage cable was played out from the cut-off Evaluation Truck to the Transmitter Truck.
By way of this cable you will test if the stages of the V_0-Test Receiver operate correctly via the Rhombus antenna. After this you will check the entire frequency apparatus.
Call the Transmitter Truck and give the following order:

"Vo-Calibrating and normal frequency Auxiliary Transmitter and Modulation engage at the First Cabinet".

A. <u>Testing the Vo-Test-Receiver at the V_0-Test Device</u>

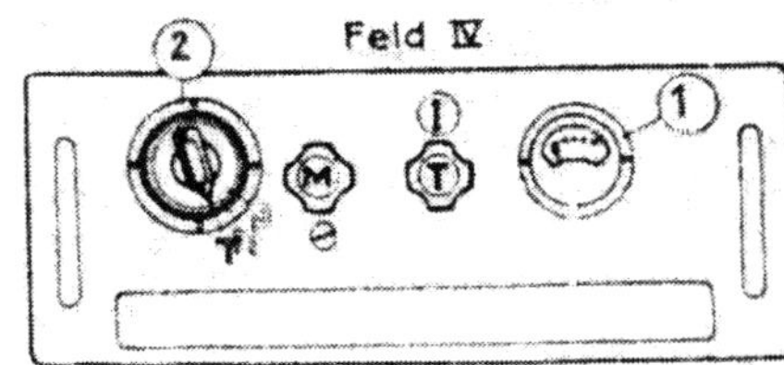

In Field I	Place Lever Switch "Reading" to left.
In Field II	Turn Button T completely to right on Position 1. Turn Test-Transformer (2) on Position O and depress Switch lever. Deflection of Pointer (1) must be within green zone.
<u>Remember</u>:	Be careful to distinguish the Test Transformer (2) with Position T and M from Buttons T and M.

	Turn Test Transformer (2) on Position M and depress Switch Lever. Turn Button M until Test Device (1) shows a maximum deflection. The deflection must be within the green Zone. Secure Button M by utilizing the screw.
	Place Test Transformer (2) on Position T and depress. By turning Button T to left the deflection of the Pointer (1) must be carried into the yellow Zone. <u>Otherwise</u>--the receiver is not modulated correctly.
In Field VIII	Place the Button P into Zero position and then turn it to left.
In Field I	a surging sound must be heard in the loudspeaker (2).
In Field III	the Deflection of the Pointer (3) must increase.
In Field IV	Place the Test Transformer (2) in position A and depress; deflection of the Pointer (1) must show approximately 1 to 3 mV. <u>Otherwise</u>--the output voltage will not be correct.
In Field VIII	Turn Button P to left until Test-Device (3) in Field II show highest reading.
In Field IV	Place Test-Transformer (2) in Position R. Watch the Deflection of the Test Device.
In Field VIII	Turn Button P from its original position to the left and then to the right.
In Field IV	The deflection of Pointer (1) must drop both times. <u>Otherwise</u>--the Automatic Control of the receiver is not correct.

B: <u>Testing the Frequency-Testing Apparatus</u>

During this test you are to follow the same proceedings you did in paragraph 1.
In Field VIII you are to slowly turn Button P from Position O to Position 10.

3. <u>Case: Testing with Vo-Calibrating Transmitter and Cut-off Signal Transmitter</u>

Tests are now conducted in order to determine if all instruments of the entire cut-off ground installatron collaborate.
Call up the Transmitter Truck and order: "Engage all devices for signal testing".
After receiving an answer from the Transmitter Truck, give the order: "Engage the Lever Switch Take-off Testing at the Switchboard of the Transmitter Truck".

In Field I Lamp "Take-off" will go out;
after approximately 56 Seconds Lamp "Transmit" will light up.
In Field VIII Turn Button P very slowly to the left.
In Field I Lamps "Reception" and "Warning Signal" and "Home Signal" must light up in succession.

Remember: Due to the lighting up of the Lamps "Warning Signal" and "Home Signal" the indicator signal 1a and 1b will appear at the Switchboard in the Transmitter Truck.
As soon as the signal 1a and 1b appear have them verified individually from the Transmitter Truck.

PREPARING THE Vo-TEST-DEVICE FOR LAUNCHING

Remember: In Field IV the V_0 test Receiver is still to be adjusted.
Call up the Transmitter Truck and remain in continuous phone connection.
In Field IV move Test-Transformer (2) in Position T and depress.
Give the following order to the Transmitter Truck: "Transmit V_0-Calibrating Transmitter for one second".

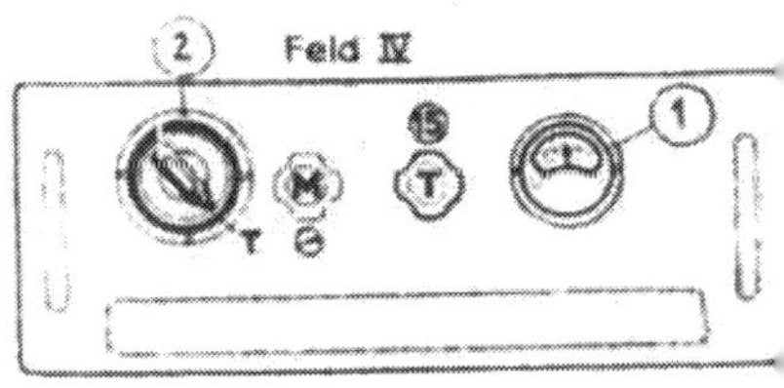

By turning Button T in Field IV you will set the deflection of Pointer (1) in the yellow Zone. The position of Button T must then lie between 5 and 20.

Very important: You should if possible arrive at this during the first transmission due to the fact that the antenna of the Vo-Calibrating Transmitter is turned on radition during testing.

The Enemy would be glad to know
the frequency you have used for transmitting
so that he could cause interference at launching.
Conduct the test therefore with utmost speed.

After Completing the Test

Report to the officer in charge that the entire cut-off ground installation is clear.
The all-clear report to the Battery Commander takes place at a specific time, for example, X—15 minutes, via phone.

Be careful: After giving the all clear to the launch control car all instruments of the Combustion Cut-off Signal site must be watched over until launching.

LAUNCHING PROCEDURE

Remember: The entire cut-off site operates completely automatic; that's why, under normal circumstances, you are not to make any adjustments.

V_0-Test-Device

After take-off of the A-4 watch the process of the VO-Test Device.

In Field I — Signal Lamp "Take-off" will go out. Immediately engage your stop-watch; after approximately 50 Seconds Signal Light "Transmit" will light up due to the contact timer in the Transmitter Truck. At the same time a rising sound must be heard in the Loudspeaker. The Test-Device in Field III will show a deflection.

In Field II — the left Test-Device will show an increasing deflection. As soon as the Zone 0.2 to 0.4 mA is surpassed the Signal Light "Reception" in Field I must light up. At the same time the pointer of the right Test-Device will deflect all the way to the left and then slowly return to O again. As soon as the Zero Position has been reached the Signal Lamp "Warning Signal" located in Field I will light up.

The Pointer of the right Test-Device will again deflect completely to the left in Field II and then slowly return to Zero Position. After returning to the Zero Position the Signal Lamp "Home Signal" in Field I will light up.

Due to this the combustion cut-off signals are initiated. The impulses have, by way of the FF-Cables, engaged the cut-off signal Modulator with transmitter in the Transmitter Truck for output and radiation of the combustion cut-off signal.

Very important: Should you notice at the V_0-Test-Device, after the Signal Light "Take-off" has gone out and the Signal Light "Transmitter" has gone on, that there is no tone in the Loudspeaker and no pointer Deflection, at the Measuring Device in Field II and III, then it is very probable that either the V_0-Calibrating-Transmitter, the Frequency Doubler or the V_0-Test-Receiver is out.

You will then by utilizing your stop-watch manually give combustion cut-off with the two push buttons in Field I.

Should there be no sound in Field I,
MORAL: Then think of the Gotz von Berlichingen,
Take a look at your stopwatch and employ
Combustion cut-off manually.

MOTTO: Flying becomes stable through steering.
The guide beam leads exactly to the target.

The electric path is produced due to a Guide-Beam Device. It is called the Guidebeam Plane.
It leads to the target from the Guide-Beam site by way of the Firing site.
The A-4 follows this path the way a hunting dog follows the trail of a wild animal.

Be careful: Erect and operate the Ground Device only if the Guide Beam is utilized for launching.

After receiving the order from the Officer in Charge park your vehicle at the point designated by the Surveying Crew.

ERECTION OF THE DIPOLE

Park the two dipoles in the designated spots.
Loosen the clamps of the Mast Bracket Stand (1) from the first Dipole.
Place Dipolehead (2) on the Mast.
Take the Dipolearms from the Containers (3) and fasten them to the Dipolehead.

(4) Set the correct atm length. You will find the specifications on the Calibration Chart.
Place the Condenser C 1 according to the Caiioration Chart on the correct value.
(5) Unfold the legs.
(6) Erect the Dipol.
(7) Disconnect the Wheel frame.
(8) Place the Dipol, together with the Groundspindle, in a horizontal Tail position. For doing this you will utilize the Clineometer.

Straighten the dipole.
Place Condenser C 2 according to the Calibration Chart on the correct value
Follow the same procedure with the second dipole.

Remember: Dipoles must be placed accurately in the designated spot and be lined up in one

LAYING THE CABLE

Lay out the Power Line Cable 220/380/50 Hz from the Generating Device to the Transmitter Truck. Lay out the High-Frequency Cable by utilizing the Tool Carrier Truck. Do not bend the Cable.
Do not lay out a bend of under 1 meter in diameter.
Connect the High Voltage Cable to the Transmitter and to both Dipoles.

OPERATING THE GENERATING DEVICE

Engage the Generating Device.
By utilizing the Handwheel (1) you will regulate the voltage at the Voltmeter (9) on 380 V.
Very Important: Frequency meter (3) must show 50 Hz.. At a Fluctuation place it on Stake (4).

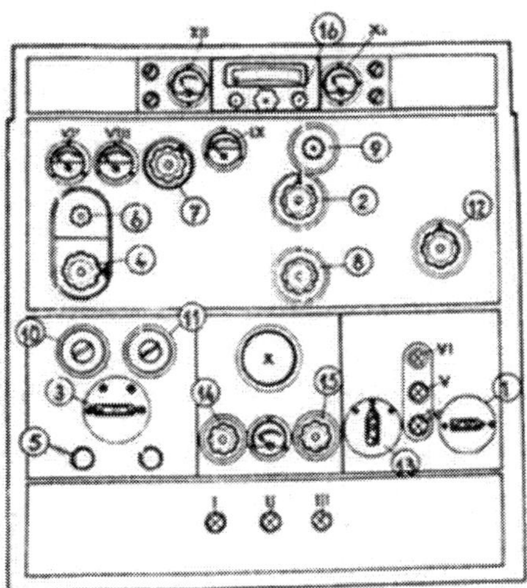

Place Switch (5) to the top.
Otherwise--the Transmitter Truck will get no power.

OPERATING THE TRANSMITTER TRUCK

Place the left Main Switch in the Transmitter Truck to the top.
Otherwise--no light will burn and the right Switch will be without power.
Place the right Main Switch inside the Transmitter Truck to the top.
Otherwise--the Rectifier Part wilt get no power.

TUNING THE TRANSMITTERS

In order to do this you will operate switches located on the Transmitter and on the Lobe-Switching Device.
On the Transmitter
Check if the Neon Lamps I, II, III are lit in the rectifier.
Place Switch (1) on turning by turning it to the left.
Turn Button (2) to left; by doing so you loosen antenna coupling.
Turn Switch (3) to right.
Otherwise--you will not disconnect Modulation.
Turn Switch (4) on the ordered Frequency.

The Lobe-Switching Device

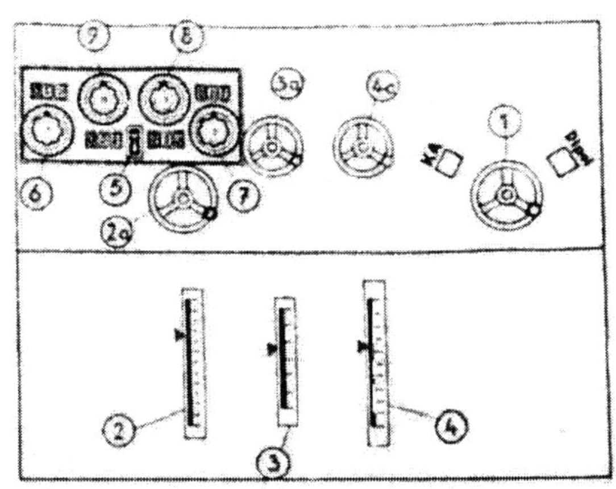

Turn Handwheel (1) on KA (mock Antenna)

Otherwise--it will be possible for the enemy to detect you.

Adjust Pointer (2) with Handwheel--(2a), Pointer (3) with Handwheel (3a) according to the Calibration Chart to the correct value. By utilizing the calibration chart set the control buttons (6), (7), (8), (9) properly and according to the ordered frequency.

You have hereby tuned the Lobe-Switching Condensers.

The Transmitter

Depress Button (5). Neon Lamp IV must light up.

Otherwise--you have no grid voltage in the tubes.

After one minute depress button (5) again

Neon lamps V and VI must light up.

Otherwise--the tubes have no anode voltage.

High frequency with trigger peaks

Hochfrequenz mit Tastspitzen

Niederfrequenz-Pakete

Low frequency packages

Turn Button (6) until Grid Current Meter VII shows a maximum value.
Turn Button (7) until Grid Current Meter VIII shows a maximum value.
Turn Button (8) until Anode Current Meter IX shows a minimum value.
Turn Button (16).

Otherwise--the Current Meter is not engaged.

Turn Button (9) until the Current Meter shows a maximum value.

At the Lobe-Switching Device

Place Switch (5) to the top. The Lobe-Switch Motor will now start up.

At the Transmitter

Turn Switch (1) to the right on Operation.
At the Oscillogram X you are now able to see the High Frequency with the triggerpeak produced by the Transmitter.
Turn Button (3) completely to the left.
At the Oscillogram X you are now able to see the Low Frequency package between the Trigger-Peaks.

Remember: Adjust the Package Width with Button (10) and (11).
The Low-Frequency packages must be located exactly between the Triggerpeaks.
At a small Deviation re-adjust the frequency at the Generating Set.
At a large Deviation re-adjust the Synchronous Condenser at the Lobe-Switching Motor.
Set the Height of the Low Frequency package at Control Button (12).

Turn Switch (13) to the left, Middle and right:
The Mismatch Indicator II may not deflect over 1,3 in the above-mentioned three Positions.

Remember: Should Test-Device XI show a larger Deflection then look for the Error in the Lobe-Switch Device.
Be sure the High-Frequency Cable has been connected correctly.

Connect with the Guide Beam Controlling Apparatus.

At the Lobe-Switching Device: Turn Handwheel (1) on Dipol: Transmitter will operate.
Turn Switch (13) to the left, Middle and right: Now again the Mismatch Indicator II may not deflect over 1.3 in the above-mentioned three Positions.
Turn Switch (13) to Middle Position.

At the Transmitter the Test Device XII is connected with the Guide Beam controlling Apparatus through the F F Cable. It will indicate if the Guide Beam plane leads to the Target by way of the Launching Site.

At the Lobe-Switching Device:
Turn Handwheel (4a) until the Pointer of the Test Device XII at the Transmitter is on Zero.
Otherwise--the A-4 will not arrive at the Target.
Turn Handwheel (1) on KA (Mock Antenna).
Conduct yout Test with utmost Speed,
for your transmitter is operating the enemy is listening in!

At the Transmitter
Turn Switch "Guide Beam clear" in the Transmitter Truck to the top. Then in the Launch Control Car the Lamp "Guide Beam Clear" will light up.

At the Lobe-Switching Device
Turn Handwheel (1) (for the time being X-1) on Dipol and check the Guide Beam Position and the Mismatch.

At the Transmitter
Depress (after 1 1/2Minutes) Button (17) and turn the Main Switch in the Transmitter Truck down.

Otherwise--your Transmitter will continue to operate unnecessarily!

MORAL: You must utilize speed during each Transmission.
Otherwise you will help the enemy to detect you.

MOTTO: At enemy jamming you are required to change your Frequency at utmost speed.

Years ago the beautiful woman of Venice changed their costumes several times during a Masquerade ball. This was done in order not to be recognized. When they re-entered the dance they not only wore a different Masquerade Costume but also matching Shoes, a matching Mask and a matching Hair-do.

If we have reason to believe that the enemy has recognized our frequency we will change it. The individual Devices must be tuned accurately when changing to a new Frequency.

The A-4 is guided by two or as the case may be three Radio Communication Systems. In order to lessen the Longitudinal deflection the following Devices are utilized:

the Vo-Testing-System
the Combustion.Cut-off Signal System.

The Guide Beam System is utilized for lessening:

the Deflection Dispersion

The following may be changed at a Frequency Change-over:

either one Radio Communication System
or two Radio Communication System
or three Radio Communication Systems.

The Frequency order is the Basis for a change in frequency. In this order:

green--V_0-Test System
red---Combustion Cut-off Signal System,
yellow-Guide Beam System.

You must differentiate between:

1. a High Frequency Change
2. a Low Frequency or as the case may be Audiofrequency Change.

The three radio communication systems contain altogether nine Frequencies which may be utilized for Transmitting. At enemy jamming one switches over to a neighboring Frequency.

Should the enemy jam this Frequency also then a Frequency located at the edge of the system will be utilized.

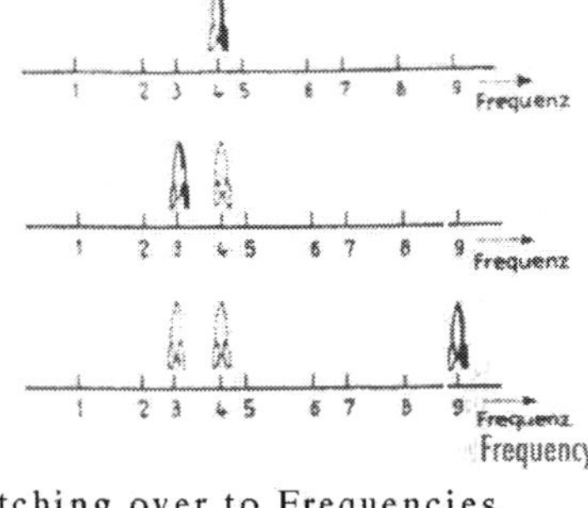

Preliminary Cut-Off Signal

Main Cut-Off Signal

Tonfolge
1
2
3
4
5

Brennschluß Vor-Kdo.

Brennschluß Haupt-Kdo.

F.. Switching over to Frequencies

there are 20 different Audio Frequencies available. The Combustion Cut-off Signal is transmitted in two Parts to the Preliminary Cut-off Signal and the Main Cut-off Signal. Each of the Signals are made up of a succession of two Double Frequencies. In order to utilize these 8 Double Frequencies as operational frequencies 5 Sound Frequencies are utilized according to a certain Schema.

The Sound Frequencies are filtered in the Cut[illegible]f Signal Receiver. Five Filters are ins[illegible]ed in each cutoff Signal Receiver. They are chosen from 15 available Filter groups.

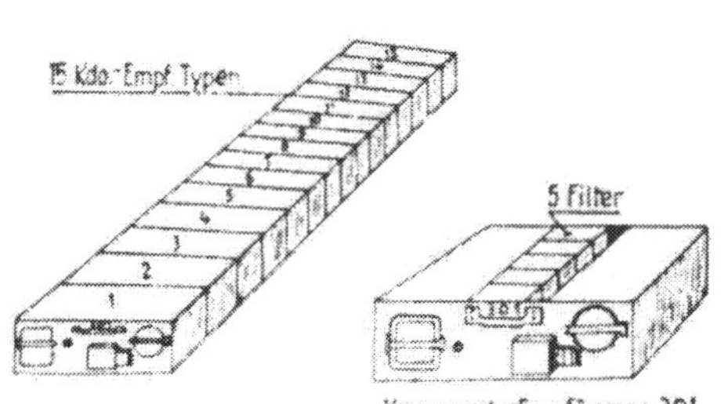

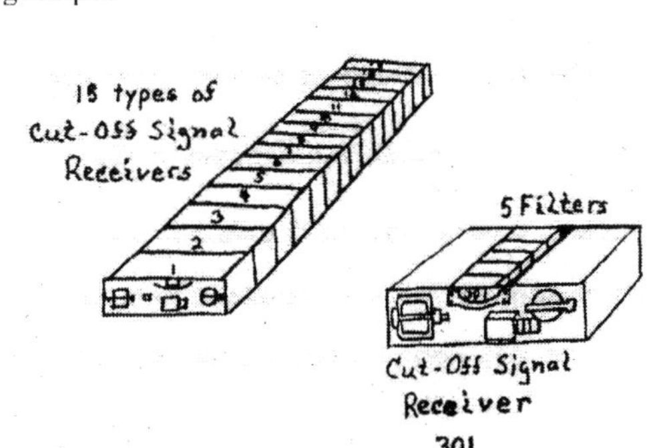

In addition it is possible to change the series of the 5 Sound Frequencies of the Signals by interchanging a Multiple Plug, the "Sequence Selector Plug".

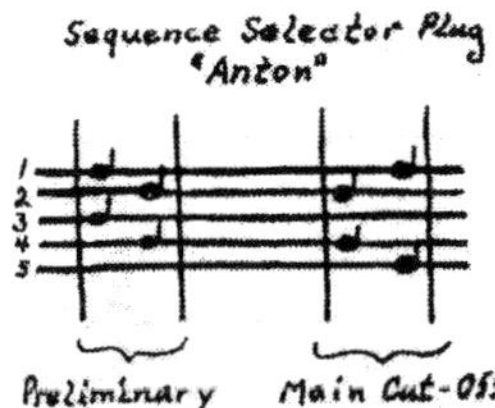

<u>Remember:</u> The Remote Control Site and the Launching Site receive the same Frequency Order.
The instruments to be adjusted will be tested individualy
in each site.

<u>Now Pay Attention:</u>

What happens if you or one of your Comrades set the wrong Frequency
in one of the two sites?
The Test will show however that the Instruments of both Sites are in Order.

But collaboration of the various instruments will not take place.
Thus no combustion cut-off can be given.
Think of the Harmonica players; one is playing in C-Major.

the other in G-Major. Each has practiced: "It's so wonderful

to be a Soldier". But there will never be any coordination

because they play in different Keys.

Example: Old Frequency Order "green 3".
New Frequency Order "green 8".e

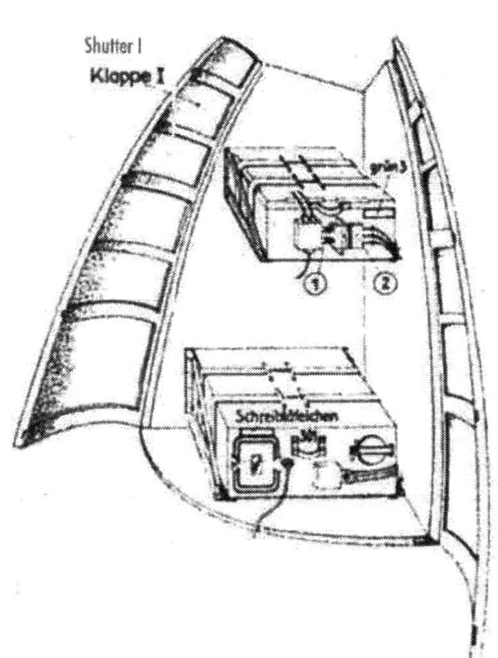

When changing the Frequency you will work with:

1. The A-4 Frequency Doubler
2. The Test Car (B-Framee) Test Transitter

1. A4-Device

You must interchange the Frequency Doubler reachable through Shutter I of the Instrument Compartment.

Unscrew the Connections (1) for the Door Antenna located at the Frequency Doubler and pull out the Plug (2).

Remove Frequency Doubler "green 3".

After receiving the Frequency Order engage the Frequency Doubler "green 8". Connect up plugs (1) and (2) again.

Tighten the Connection Nuts of Plug (1) for Antenna connections by hand only and not with a pair of pliers.

2. Test Car (B-Frame)

Instead of Test Transmitter "green 3" install Test Transmitter "green 8" in Compartment 4.

You will find it in Field 14 to 21.

Now you will test the new Frequency Doubler.

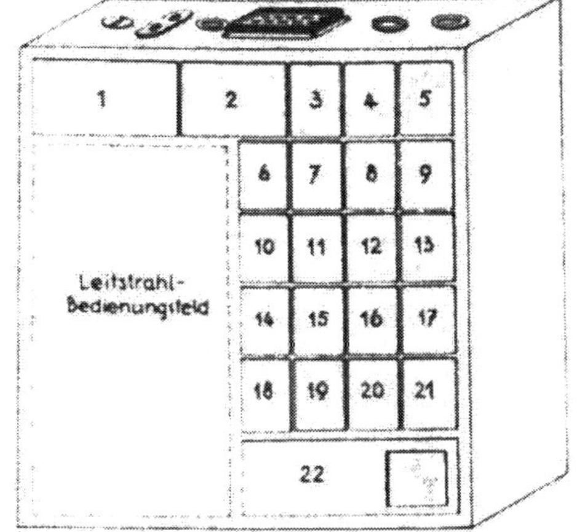

II. COMBUSTION CUT-OFF SIGNAL SERVICE

Example:

Old Frequency Order: "red 6"; 301; Anton.

New Frequency Order: "red 4"; 301; Emil

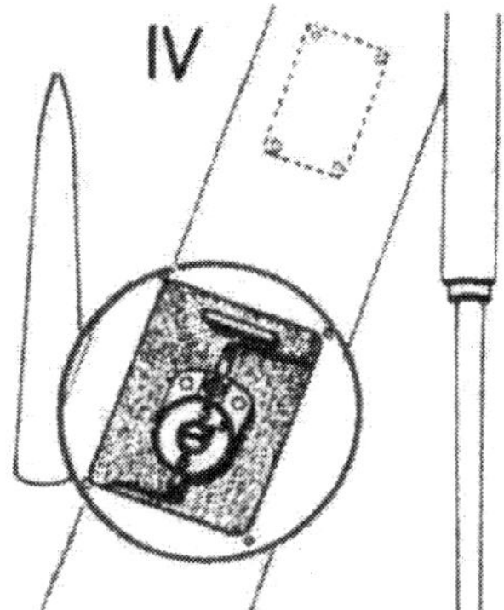

There is a change of Carrier Frequency as well as of the 5 Sound Frequencies. Interchanging the 301 Group is hardly necessary.

It would mean interchanging the entire Cut-off Signal Receiver.

When changing the Frequency you will work with:

1. A 4: Cut-off Receiver and Antenna belonging to Fin 2 and 4.
2. Test Car (B-Frame): Test Transmitter, Cut-off Signal.

1. A 4.

At the Cut-off Receiver, reachable through Shutter 1, you are to interchange the following mechanisms or Plugs:

Open the left Shutter to the Cut-off Signal Receiver.
Pull out the High Frequency Plug "red 3" and after receiving the order insert Plug "red 4" and close the Shutter well.
Open the right round Shutter.
Exchange the inserted Sequence plug "Anton" for "Emil" and close the Shutter well again.
Mark down the change on the Writing Tablet provided for this purpose.

Beside this you are to re-set the Antenna on Fin 2 and 4.
Open the lower Antenna adjustment box.

On the Trimming Condenser there are 3 Plates marked with 2. 5 and 8.

For the Frequencies red 1 to 3 you will adjust Plate 2.
For the frequencies red 4 to 6 you will adjust Plate 5.
For the frequencies red 7 to 9 you will adjust Plate 8.

Therefore after receiving the frequency Order "red 4" set Plate "5" on the red Marker.

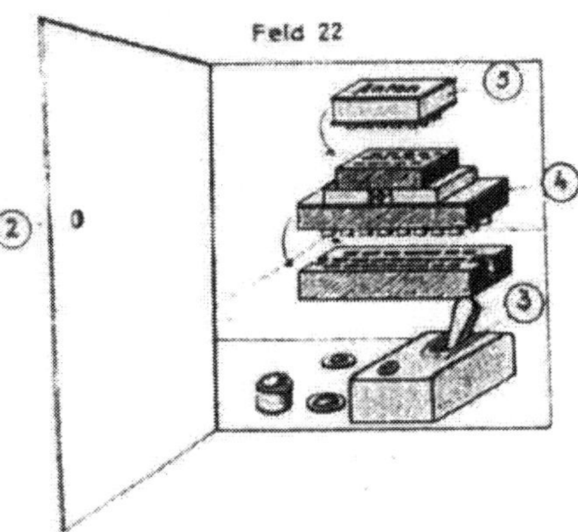

2. Test Car (B-Frame)

Install Test Transmitter "red 4" instead of Test Transmitter "red 6" in Compartment 3.

You will find it in Field 6 to 13.

Open the Shutter in Field 22 and interchange the inserted Sequence Selector Plug "Anton" for "Emil".

Now you will test the cut-off Signal Receiver as previously described.

III. GUIDE BEAM SERVICE

Example: Old Frequency Order: "yellow 3"
New Frequency Order: "yellow 4"

When changing the Frequency you will work on the following:

1. A 4: Guide Beam Device, Aircraft Antenna.

2. Test Car (B-Frame): Model Transmitter

3. Controling Apparatus

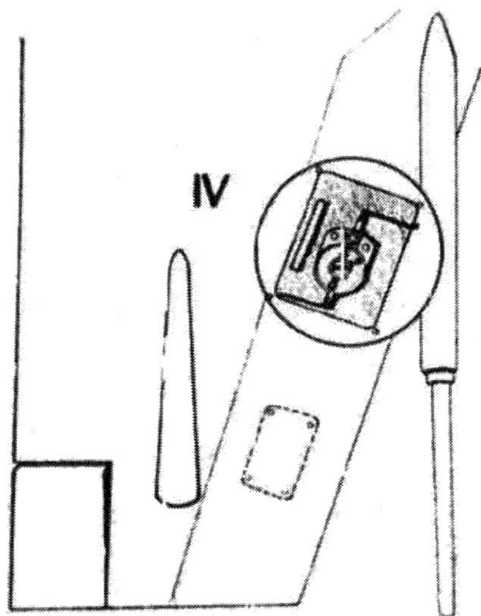

1. A 4:

You are required to interchange the Guide Beam Device attainable through Shutter IV of the Instrument Compartment.

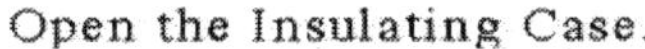

Open the Insulating Case.

Pull out Plug (1) and (2) and remove the Guide Beam Device "yellow 3".

After receiving the Frequency Order install a Guide Beam Device "yellow 4".

Insert the Plugs (1) and (2) again.

Close the Insulating Case well again.

Reset the Dipole Antenna at Fin 2 and 4.
Open the upper Antenna Tuning Box. On the Trimming Condenser there are 3 Plates with the Markings 2, 5 and 8.

For the Frequency "yellow 1 to 3" adjust Plate 2.

For the Frequency "yellow 4 to 6" adjust Plate 5.

For the Frequency "yellow 7 to 9" adjust Plate 8.

After receiving the Frequency Order "yellow 4" set Plate 5 on the yellow Marker.

2. Test Car (B-Frame)

Instead of the Test Transmitter "yellow 3" install the Test transmitter "yellow 4" in the Model Transmitter. You will find it in one of the fields under the Model Transmitter.

Now you must test the Guide Beam Device, as previously described.

3. Signal Apparatus

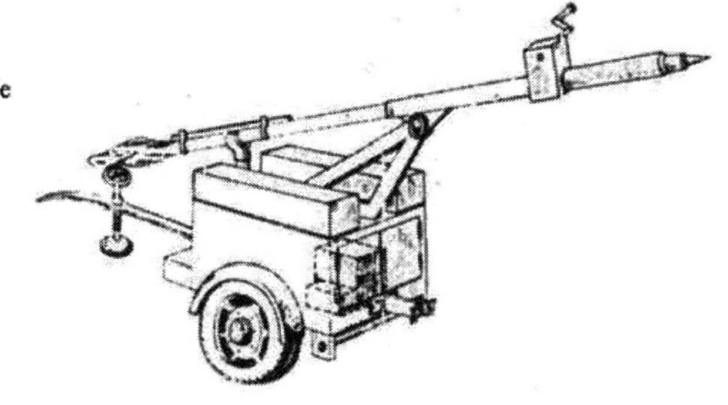

In the Signal Apparatus you are required to exchange the Receiver and the High Frequency Amplifier.

Open the left Shutter of the Trailer.

Instead of the Receiver

"yellow 3 install Receiver "yellow 4",

instead of the High Voltage Amplifier "yellow 3" install the High Voltage Amplifier "yellow 4" in the Signal Apparatus.

After switching these Devices test the Signal Apparatus again.

1. V_0-Test-Service

Example:: Old Frequency Order:
"green 3"; ABC-425; D-1; E-2-0.
New Frequency Order:
"green 8"; ABC-449; D-2; E-3-0.

When changing the Frequency you will work with:

1. Transmitter Truck: Vo-Test-Transmitter with normal Frequency Auxiliary Transmitter.
2. Cut-off Evaluation Truck: V_0-Test-Frame.

1. The Transmitter Truck

You are to set the new Carrier Frequency "green 8" at the Vo-Test Transmitter and normal Frequency Auxiliary Transmitter.

At the First Cabinet of the Vn-Test-Transmitters

In Field V Place the lever of the "Frequency-Adjustment 1" from Position 3 to Position 8. Quarz Nr. 8 is hereby engaged.

In Field I & III Turn the Knob "Frequency-Adjustment 2 to 5 on Position 8.

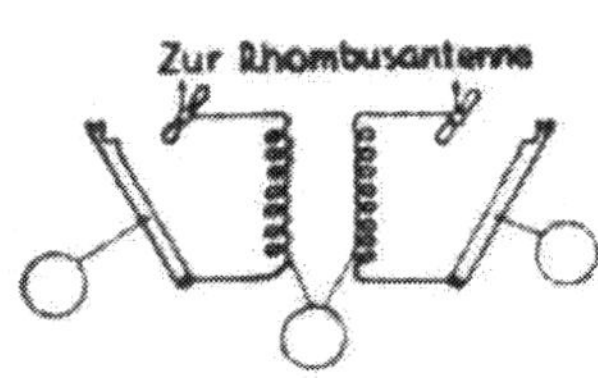

At the Output Cabinet of the V_o-Test Transmitters

Open or shut both Lashes (1) of the two coupling Soils (2) on the Output Cabinet.
Just how to set the Lashes for the Frequency "green 8" is described in the plate above it.
After this adjustment the Transmitter must be retuned.

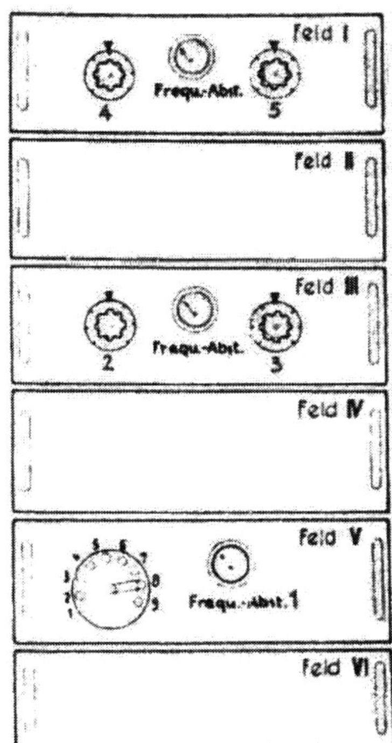

2. Cut-off Evaluation Truck

You are to adjust the new Frequencies at the V_o-Test-Frame.

Adjustment of the New Resonance Zone E 3-0.

In Field III Turn Switch 1 on Position 3
Turn Switch 2 on Position 0

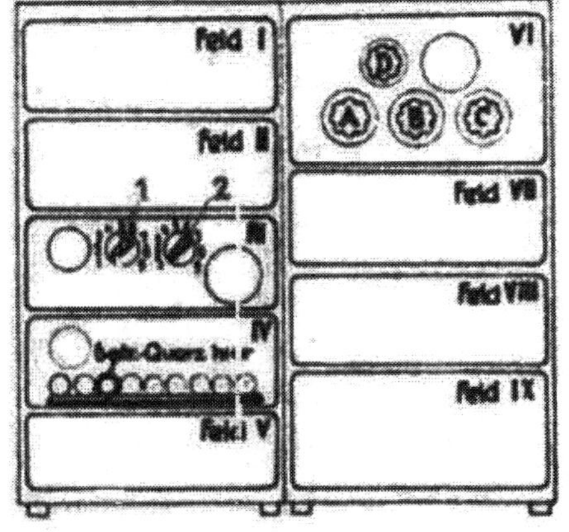

Adjustment of the New Oscillator Stage "green 8"

In Field IV Open the Shutter and place the Quarz "Po 8" in white-edged set of Sockets with the Description "Operationa Quary here".
Close the Shutter again.

Adjusting the New Launching frequency ABC-449 and D-2

In Field VI Set Test Switch A on 400
Test Switch B on 40
Test Switch C on 9
Turn Button D on Position 2

After these Frequency Adjustments you are to conduct all tests for the V_o-Test-Frame.

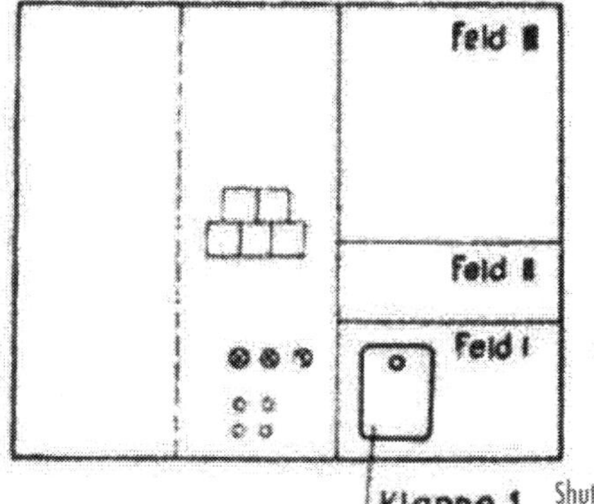

II. COMBUSTION CUT-OFF SIGNAL SERVICE

Example- Old Frequency Order: "red 6"; 301; Anton
New Frequency Order: "red 4"; 301; Emil.

When changing the Frequency you must work with the following:

1. Transmitter Truck: Cut-off Signal Transmitter
   Cut-off Signal Modulator
   Signal Control Receiver
   Phantom Signal Modulator
2. Dipolantenna

1. Transmitter Truck

At the Cut-off Signal Transmitter you will set the new Carrier Frequency.
In Field I open Shutter 1.

Set the Rotary Switch D 1 located behind it according to the amount of Quarz.

Quarz red	Rotary Switch-Position D 1
1	A
2 or 3	B
4	C
5 or 6	D
7	E
8 or 9	F

According to the Frequency Order "red 4" Rotary Switch D 1 is set on Position "C".

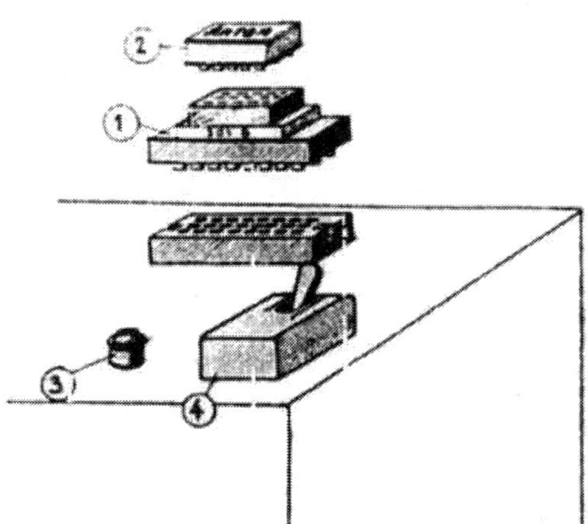

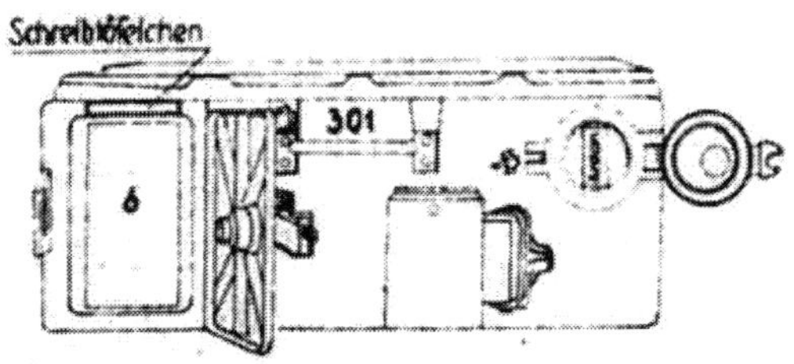

At the Cut-off Signal Modulator

Instead of the Sequence Selector Plug "Anton" located at the Frequency Group Selector Plug 301 insert the Sequence Selector Plug "Emil".

At the Signal Control Receiver

Instead of inserting Mechanism Nr. 6 insert Mechanism Nr. 4 into the left opening.

Instead of inserting the sequence Selector Plug "Anton" insert the Sequence Selector Plug "Emil" into the right round opening.

A number and a letter is marked on the Frequency Group Selector Plug which is plugged to the Cut-off Signal Modulator, in our case, for example, 301 B.

If the Frequency Group Selector Plug at the Cut-off Signal Modulator is changed then watch for the identification letter. You are required to plug the Switch with the identical identification letter to the Phantom signal Modulator.

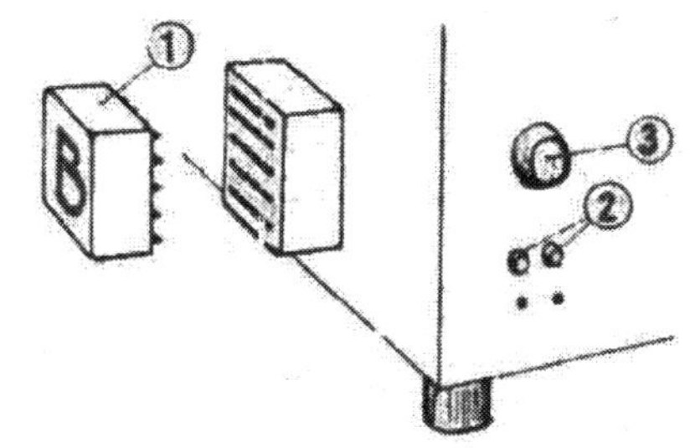

Similar to the example shown concerning a Frequency Change, you are not to make any changes on the Phantom Signal Modulator, due to the fact that the frequency Group Selector Plug 301 B at the Cut-off Signal Modulator has remained the same.

You are required to set the new values at the Dipol in accordance with an available chart giving the new Frequency.
For "red 4" the chart will give you the following:
$1 = 29$; $C = 10$; $C_2 = 155$

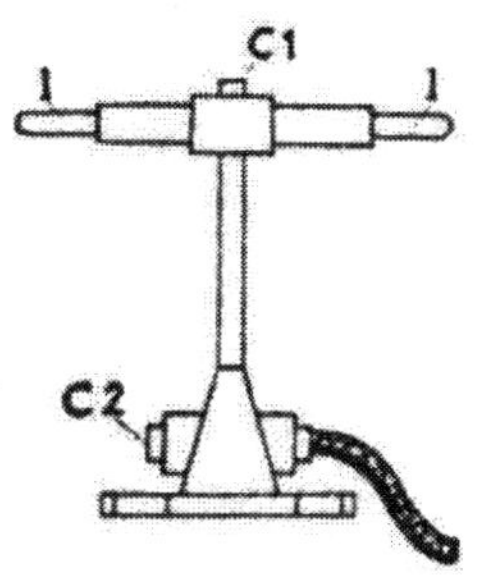

III. GUIDE BEAM SERVICE

Example: Old Frequency Order: "yellow 3"
New Frequency Order: "yellow 4".

At frequency change you will work with:

1. Transmitter Truck. Transmitte Lobe-Swi ing Devic(
2. Dipol

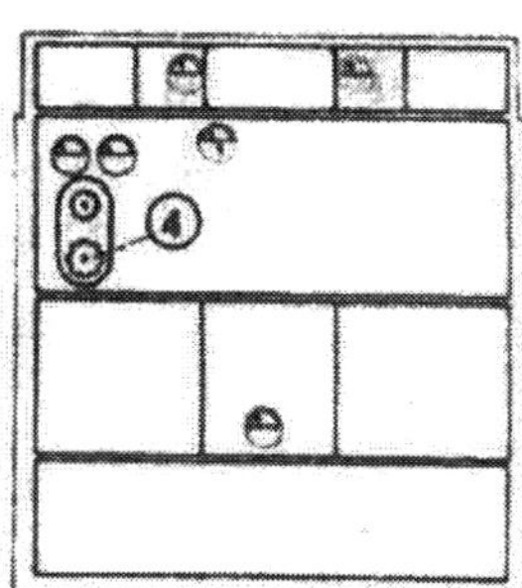

On the Transmitter you are required to set the new Carrier Frequency, "yellow 4".

Turn Switch (4) from Position "3" to Position "4".

Quarz "yellow 4" is hereby connected.

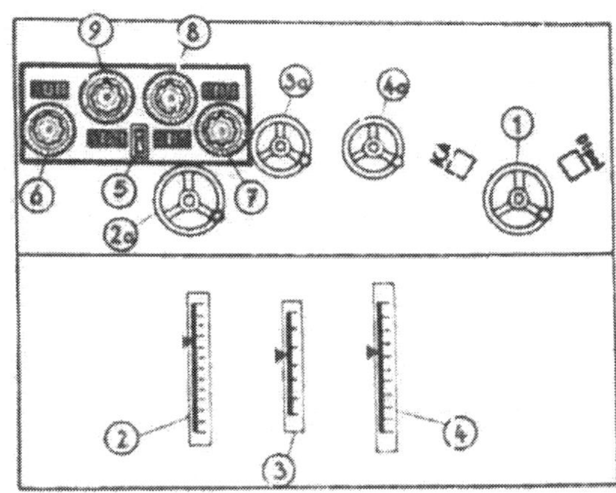

Should instead of "yellow 1 to 4" another frequency be ordered (between "yellow 5 and 9") then you must first of all install the corresponding oscillator control crystal. This is done in the following way:

Unsnap the HF-Pan.
Open the Thermostat.
Remove one crystal and put the required crystal in its place.
After that close the Thermostat, put the HF-Pan back in its place and return the Transmitter.

At the Lobe-Switching Device

Turn Buttons (6) (7) (8) and (9) on the cut-off co-efficients of the ordered Frequency. You will find these values on the Calibration Chart.

Turn Handwheel (2a) and (3a) until the Pointers are set on the cutoff co-efficients of the ordered Frequency on the lobe-switching device.

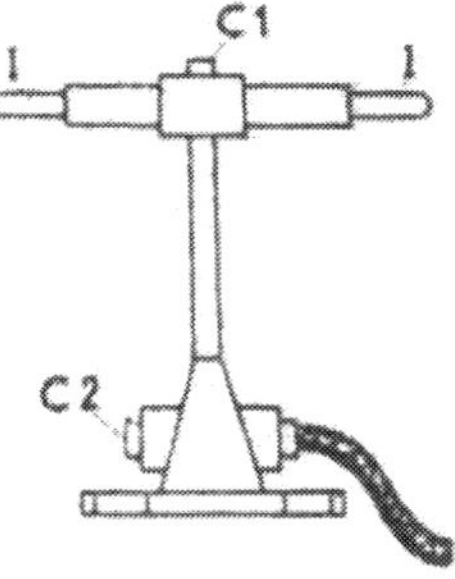

2. Dipole

Set the Dipole Arm Length 1, the Condensers Ci and C2 on the Cut-off Co-efficients of the ordered Frequency

You will find these values on the Calibration Chart attached to the Dipole.

After this adjustment you are required to tune the Transmitter again.

1. Signal Functioning in the Combustion Cut-off Site

Example Nr.	Type of Frequency-Change	Frequency Order		Dipol Adjustment	All Instruments in the Transmitter Truck						
					Cut-off Signal Transmitter	Cut-off Signal Modulator		Phantom Signal Modulate Plug	Cut-off Test Receiver		
		Old	New			Freq. Group Selecter Plug	Seq. Selecter Plug		Receiver	HF Mechanism	Seq. SelecterPlug
1	HF.	red "5"	red "7"	According to calculated Values for 1, C_1, C_2	Switch D_1 on Quarz "E"	—	—	—	Do not exchange	Nr. 7	—
2	NF-Signal	red "7" 101 A Bertha	red "7" 101 A Konrad	—	—	—	Konrad	—	Do not exchange	—	Konrad
3	NF.-Group	red "7" 101 A Konrad	red "7" 306 B Konrad	—	—	306 B	—	B	Marking 306	—	—
4	NF.- Signal and Group	red "7" 306 B Konrad	red "7" 501 A Anton	—	—	501 A	Anton	A	Marking 501	—	Anton
5	HF.- with 2 NF.	red "7" 501 A Anton	red "4" 315 F Dora	According to calculated val. for 1, C_1 C_2	Switch D_1 on Quarz "C"	315 F	Dora	F	Marking 315	Nr. 4	Dora

2. V_O-Test Service in the combustion cut-off site

Example Number	Type of Frequency Change	Frequency order		Transmitter Truck		Cut-off evaluation truck
		Old	New	First Cabinet	Output Cabinet	V_o-Test Frame
1	High Frequency	" ' green6 ABC:530 D:7 E:7–0	"" green 2 ABC:502 B:6 E:6–0	Quartz Field V Selector Plug Frequency-Adjst: on 2 Field III Frequency Adjuster Button 2 and 3: on 2 Field I Frequency Adjuster Button 4 and 5: on 2	Rhombus Antenna Join the spools open or close	Field III: Switch "E" to the left on "6" Switch "E" right on "8" Field IV: Place Quartz Po2 in two sockets "Operational Quartz here" Field VI: Control Button "A" on 500 Control Button "B" on 0 Control Button "C" on 2 Control Button "D" on 6

In order for you not to forget,
MORAL: The things you have to do,
Quickly utilize these charts
Then you will not mess uo a thing.

3. COMBUSTION CUT-OFF SIGNAL at the LAUNCHING SITE

Example Nr.	Type of Frequency-Change	Frequency Order		Test Car: B-Frame			A 4: Instrument Comp. Shutter 1			
		Old	New	Test Transmitter red	Cut-off Signal Modulator		Cut-off Signal Receiver			Antenna-tuning-box
					Frequency group SelecterPl.	Sequence Selecter Plug	Receiver	HF Mechan.	Sequen. Selecter Plug	
1	High Frequency	red "5"	red "7"	Exchange Nr. 7 for 5	—	—	(do not exahange)	Nr. "7"	—	"8" on red mark
2	Low Freq.-Signal	red "7" 101A Bertha	red "7" 101A Konrad	—	—	Konrad	(do not exchange)	—	Konrad	—
3	Low Freq.-Group	red "7" 101A Konrad	red "7" 306B Konrad	—	306 B	—	306	—	—	—
4	Low Freq.-Signal and Group	red "7" 306 B Konrad	red "7" 501 A Anton	—	501 A	Anton	501	—	Anton	—
5	High Freq.-with two low Frequencies	red "7" 501 A Anton	red "4" 315 F Dora	Exchange Nr. 4 for 7	315 F	Dora	315	Nr. "4"	Dora	"5" on red mark

4. V_0-TEST-SERVICE at the LAUNCHING SITE

Example Nr.	Type of Frequency-Change	Frequency Order		Test Car: B-Frame	A 4: Instrument Compartment Shutter 1	
		Old	New	Test Transmitter green	Frequency Doubler	Remarks
1	High, Frequency	green "6"	green "2"	Nr. 2 with 6 Interchange	Entire Device Interchange. Reinstall gr."2"	No Door Antenna tuning necessary

5. Guide Beam Service at the LS-Site

Example Nr.	Type of Frequency-Change	Frequency Order		Apparatus in the Transmitter Truck		Dipol-Adjustment
		Old	New	Transmitter	Lobe Switching Device	
1	High Frequency	yellow "3"	yellow "4"	Switch (4) on "4"	Adjust the Buttons (6) (7) (8) (9) and the Handwheel 2a,3a according to chart	According to chart Values for 1, C_1,C_2

6. GUIDE BEAM SERVICE at the LAUNCHING SITE

Example Nr.	Type of Frequency-Change	Frequency Order		Test Car: B-Frame	A 4: InstrumentCompartment Shutter IV		Control Site	
		Old	New	Test transmitter yellow	L.S-Device	Antenna-tuning-box	Ultra short wave Receiver	Amplifier
1	High Frequency	yellow "3"	yellow "4"	Nr. 3 for 4 Interchange	Nr. 3 for 4 Interchange	"4" on yellow mark	Nr. 3 for 4 Interchange	Nr. 3 for 4 Interchange

NOW AVAILABLE!

ISBN #978-1-937684-76-1

37095429R00098

Made in the USA
Middletown, DE
19 November 2016